ATOMIC SPIRITUALITY

FAITH ON A QUANTUM PLANE

Dylan Clearfield

Copyright © 2017 by Prism Thomas

G. Stempien Publishing Company
All rights reserved

ISBN 978-0-930472-19-1

Acknowledgements

Sincere thanks to Daniel Wilson, acoustical-engineer and author, for the invaluable information about the private life of James Bathurst (John Palfrey) and his deep insights into the book *Atomic-Consciousness*. Also appreciated is his eye for corrections.

I also express deep gratitude to Wendy Thomas for her many hours of verbal lessons concerning the modern spiritual belief system, particularly for her introduction to Paramahansa Yogananda.

I of course alone am responsible for any errors that might be found within the forthcoming pages.

CONTENTS

BEING HERE

PART I – SPIRITUAL OPERATING SYSTEM

IS THERE ANY REALITY?
    Reverse reality
    The Babel effect
    Law of ultimate potentiality

ATOMIC-CONSCIOUSNESS
    James Bathurst

INFINITE PRESENCE AS STATED
    What does conscious mean?
    Everything is conscious
    Types of consciousness
    Unified-consciousness

EVIDENCE OF INFINITE PRESENCE
    Non-causality
    Wish fulfillment
    Law of attraction
    Negation
    Negative realities
    Material precognition

CONTACTING INFINITE PRESENCE
    The neutral mind
    Electro-magnetic contact
    Direct contact – meditation

RECURRENCE AND SPIRITUAL RENEWAL
- Re-enactment
- Replay effect
- Mathematics of recurrence
- Eternal recurrence
- Apocatastasis
- Akashic records/eternal memory
- Appointed time of death

PART II – MULTI-VERSE EXISTENCES

NANO SCIENCE AND ANGELS
- Weighing angels
- Angels of the periodic table

SCIENCE OF THE APOCALYPSE
- Newton calculates the End Times
- Prophets of the Bible
- Three Days' Darkness
- February 4, 1962 – planetary alignment
- Doomsday clock
- Mayan calendar
- The Bible code

QUANTUM IMAGING
- Visions
- Magical sights

PROPHECIES AND PREMONITIONS
- Daily mind miracles
- Harmonics from the future
- Voices from spirits
- Mechanical consciousness
- Combined imagery

ALTERNATIVE EXISTENCES
    Possibility factor
    Déjà vu clue

TIMELINE OF ETERNITY
    Many worlds of Hugh Everett
    Timelines of lives
    Creating an alternative reality
    Reverse Déjà vu

ALTERNATIVE AFTERLIVES
    Prime theory
    Anthropological assessment

APPENDIX 1
Curious observations

APPENDIX 2
Excerpt from Voltaire's "Micromegas."

Bibliography

Being here

This book presents an alternative concept to the currently prevailing ideas of spiritual belief, philosophy and science. It is not meant to be totally original and it includes acceptance of the basic ideas as proposed by such major figures as: Paramahansa Yogananda, Sri Nisargatta Maharaj, Maharishi Mahesh Yogi (transcendental deep meditation) Eckhart Tolle, Edgar Cayce, Gary Weber, Gregg Braden, and others who profess similar primary viewpoints.

The model that will be proposed concerning the ultimate destination of all of reality and everything that is part of it will be of a unique nature. This involves experiencing all potential possible existences which will be finally coalesced into a singular point of Being by the Eternal Observer. This will be fully demonstrated later. For the scientists: this theory concerns the application of the ultimate measurement which will make all of reality observable in one location! It is a quantum study of eternity and how it will be experienced by the individual.

For those who are not familiar with quantum physics, it should be noted that making detailed measurements is critical to this field of study. In many instances, it is the measurement that is taken of a phenomenon that gives it reality and even its very existence. This confers upon the role of "Observer" a special importance, no matter if the observer is human, or higher intelligence, or God.

Concerning current belief systems, the concept of non–duality is accepted, herein referred to as the state of everything being one thing (as per musician and philosopher Ted Myers, among many others). The role of the ego as a phenomenon which is desperately trying to control our lives by placing those "blah, blah, blah" sounds in our heads (as per Gary Weber) is also acknowledged here. And the general concept that the universe is a process that is unfolding and that we are taking part in it as minor observers with little control and restricted free will is also part of this book's belief system.

Among the scientists who will be referred to are the following. Niels Bohr (so–called father of quantum physics and the Copenhagen interpretation), Sir Isaac Newton, Erwin Schrodinger (ala Schrodinger's cat), David Deutsch, Bryce DeWitt and, of course, Hugh Everett, originator of the Many Worlds Theory of quantum physics.

The main departure point from the spiritual concepts will be in the scope of reality considered here (as in the existence of an infinite multi–verse). Also significant is the overall influence of consciousness as it occurs in everything that exists. And of prime importance is the power of potentialities to the effect that they are in themselves true realities and that they must exist on all levels. To this end the many worlds Theory of Hugh Everett will be used as the model on which this book's main premise is indebted. All realities must follow every potential alternative possibility to attain that singularity of Being which is the ultimate source. Infinite Presence is the Essence of being, or non–being and, according to this book, is a conscious, benevolent entity.

This work will rely heavily on scientific concepts and scientific proofs to convey its major points. However, it will repeatedly be demonstrated how these scientific concepts reflect, if not copy exactly, the viewpoints espoused by most major religions and spiritual belief systems. Only the wording is different. The meaning is the same. The hope is that people who claim to be atheists or are otherwise non–believers might recognize this convergence of science with faith to form a deeper understanding of how the two might be the same thing.

The type of God, or Infinite Presence, that will be defined here can be seen as an impersonal, uncaring intelligence if one wishes; however, it will be demonstrated throughout this work that a consciousness is in control of the effects that arise and that its purpose is directed and with benevolent intention. The reader will be free to deny this interpretation. This book does not promote its own dogma. But it does have a basic theme which you can reject, consider, or – maybe – even accept.

For those who object to the scientific nature of this work, I remind them that the quantum theory upon which the foundation for this book rests is a manifestation of the Creator. It could not exist without his direction. It can be verified by experiment. Why would this not also be a valid part of his creation?

The basic strategy of this book is to examine earthly activities, both commonplace and extra–ordinary (spiritual), and compare them with the quantum theories propounded within and, in comparing these phenomena, devise a link between them and the conscious actions of the supreme spiritual entity. Then to determine whether there is benevolence associated with this entity, imbuing it with what would be called divine attributes

– God. Finally, to examine the everlasting nature of existence in a way comprehensible to the human mind.

In addition, considering the concept that all things are conscious to various degrees, examples will be provided to show how these various levels of consciousness contact one another as well as the Infinite Presence. While this idea has been discussed before in general terms by other researchers, this work will deal in specifics and reveal precisely how, when, and sometimes why, these different levels of consciousness interact on an atomic scale. This will be examined on both the scientific and spiritual plane with the intent being to demonstrate how they are the same concept differing only by the language used by which the events that take place are described.

One of the most difficult, yet critical, concepts to be presented and demonstrated is the idea that an effect of an outcome can often precede its own cause. This is counter to all laws of Causality that have been held to be unalterable. Events can and do take place before they occur. In religious as well as scientific terms, this would explain premonitions and prophecies. The being that is called God had already existed before coming into being and in that sense caused his own existence. Examples of these events will be given in the upcoming pages rather than mere theory.

A plea to those who are interested in this specific topic of effect preceding cause: please, do not read this book before obtaining a copy of it.

What this book will strive to show is how The Infinite Presence uses its consciousness to direct the lower levels of consciousness to affect every- day action for beneficial purposes unless altered by the observer whose imposition may produce negative outcomes, thus misery and seemingly evil results.

As already noted, the primary model around which this work revolves is the Many Worlds Theory of quantum physics created and developed by the brilliant physicist Hugh Everett. However, it will not be dogmatically followed and there will be alterations added as required. Also of significant importance to this work is the book *Atomic–Consciousness* written by James Bathurst in 1892. Its primary contribution will be in the form of evidence provided of the atomic consciousness of all matter in all universes and the attending theories.

The hope is to demonstrate how the concept of near infinite parallel worlds excellently defines the spiritual existence of all realities, leading to one ultimate essence as observed by the Infinite Presence (God). This will be presented in detail in part 2 of this work. Part 1 will involve basic spiritual concepts that will be considered in conjunction with scientific findings and thereby be combined into a unified system.

I ask for the reader's patience regarding the seemingly analytical style of the writing of this book. It is a difficult subject to describe without becoming technical and my background is in the field of science. By training I am an archaeologist, acquiring my BA degree from the University of New Mexico and continuing onward to receive upper level degree in science at Southern Illinois University – Carbondale.

PART I
SPIRITUAL OPERATING SYSTEM

Is there any reality?

REVERSE REALITY

Is there any reality? That depends on the observer. Each one of us sees what we think is a reality. It is a reality of our own. And not everyone sees the same reality. In fact, some of us only have glimpses of the other reality that other people see.

Gibberish? Writing in riddles meant to confuse? No, neither. As you read this book you will quickly find that not only will statements be made that seem to be outrageous – even insane – you will also quickly find that some measure of proof will be supplied to support these apparently outlandish observations. Such will be the case now.

Absolute proof that each of us sees a different reality was provided on February 27, 2015. An event took place – of seemingly the most trivial nature – that shocked millions of people into sudden awareness. And please note the apparent triviality of the event itself because time and again it will be demonstrated in this book that what may seem trivial may be of astounding significance.

What happened on February 27, 2015 that was so earth shattering? A woman in Scotland purchased a dress to wear to a friend's wedding.

One day without warning, the internet was set aflame by discussions concerning a question over which millions of people had a viewpoint. What was the true color of a dress that a woman bought to attend her friend's wedding in Scotland! That's right. If you do not recall this famous dress incident, it involves a piece of attire in two primary shades that were visible to various viewers. Keep in mind that these are only TWO of other alternative colors, but 99% of viewers saw either a golden and white dress or a black and blue dress.

These were not two different dresses. It was the same dress. But some people saw it colored golden and white and other blue and black. Remember the incident? What color did you see? At the time of this internet sensation, there was even a small minority of people who saw even a different color combination from the two already listed.

I performed a scientific experiment on the photographs of this dress and reversed the colors on my computer screen. When the colors were reversed, the gold and yellow dress looked blue and black. And the blue and black dress became gold and white. Does that mean that some people see reality in reverse? Which people though? What is the true color of the dress? Maybe it's neither.

People could not comprehend how someone else was seeing the exact same object as they were, but seeing it in a completely different way. It's as if you saw a stop sign painted red and the person beside you saw it painted blue. You couldn't both be right – could you?

In the case of the mystery dress both sides were right. This dress genuinely was at the same time being seen by people as being of two different colors. The same dress, seen at the same time. But this is impossible. Isn't it? Apparently, it wasn't impossible since it was being reported that this is the way millions of people were seeing the dress.

Then the scientific explanations came. Most of the "scientific" views classed the phenomenon as a visual anomaly rather than a true distortion of reality. Those who argued thus believed that some people saw the dress as blue and black because the number of color sensitive cones in their eyes were different from other peoples'. Or people who were more fatigued than others saw a different colored dress due to the fatigue factor. Not too many arguments were made in favor of the one that simply stated that sometimes reality is truly different for different people.

THE BABEL EFFECT

Most people are aware of the biblical story about the Tower of Babel and how causing the confusion of languages made interaction among the people involved impossible (Genesis 11: 1-9). Consider this regarding how each person today is living in his own reality. How is it possible for anyone to communicate with anyone else? We are all essentially speaking a different language in our thoughts as well as verbally, and perceiving the world in conflicting ways.

Therefore, it is vital that all elements of reality be possessed of some form of consciousness. This allows for direct contact to be made with each individual human mind, as well as the minds – or thought controlling systems - of any other being or object. It is a way that God, or Infinite Presence, or the First Mover Principle of Aristotle delegates the process of communication dispersal among all points of existence.

There is also another biblical parallel to note. This concerns the role of the Holy Ghost in the Christian religion and how by its intercession the original Apostles were give the power to speak in "tongues" and, in more general terms, dispensed the ability to reason beyond normal understanding. Speaking in tongues is obviously a form of communication of a more specific nature – verbalized language – but the purpose was the same as spreading consciousness throughout existence insofar as allowing for the rapid exchange of understanding among people who ordinarily would not be able to comprehend one another. And, speaking in tongues, also relies upon a very special underlying consciousness.

LAW OF ULTIMATE POTENTIALITY

There is a multi–verse of parallel worlds in existence where duplicates of our own selves are now residing. They were placed in this other world by choices, options that we took in our original timeline – no one knows which one is the original timeline – choices which once made were required to be fulfilled. It is called here the law of Ultimate Potentiality, which means that something which has a possibility of existing must be experienced to the fullest otherwise the totality of reality could not be complete.

Why is it so important that we are now existing in our own personal reality, even though interacting with all other realities? Because it is by our individual actions in our own individual reality that the choices or options are taken which creates our alternate, parallel world which contains our other alternate selves (this follows the ideas set forth in the Many Worlds Theory of Hugh Everett). This also is the eternal environment in which we currently exist.

While it may be true that that everything is illusory, it may also be true that the Eternal's desire is that the extensions of Itself – us and the rest of the universe – are to be allowed the choice of experiencing these effects to experience them as we do, as illusions, for his own purposes. Maybe this is a free choice granted all of us to accept or deny, even if it may cause personal pain along the way. The Eternal is interested in the existence and experiencing of all options.

Ultimately, the belief that you the individual personally accepts as true is the only one that matters to you. We each create and exist in our own separate reality – as demonstrated in the example of the two differently colored dresses whose appearance was different to different people. By our own choices we set into action the life of alternative worlds under the direction of the Supreme Power which allows us to determine our own fates in a limited way.

Maybe time also is an illusion as many people believe. But even so it exists in some realm and can be experienced as it is and as we choose to view it. And this includes the existence of memory, even eternal memory as will be examined in the upcoming pages.

Quantum physics reveals worlds of non–reality which in everyday understanding cannot exist yet appear to do so. It presents a factually existing level of seemingly illusory reality which operates by its own seemingly artificial rules and laws which distracts the observer into creating whatever reality is projected before him. As such, the effects of quantum physics merges with the spiritual world on another plane where what is called reality is an illusion and where the observer is the ultimate judge. But who is observing the observer?

But keep this in mind. If everything that exists was not conscious there would not be any way to distinguish reality.

Ironically, two of the foremost minds on each subject – quantum physics and the spiritual path to awareness – have offered almost the same comments about each field of study. Niels Bohr said of quantum physics, "Anyone who is not shocked by quantum theory has not understood it." And in the same vein, Maharaj remarked that anyone who claims to know what enlightenment is does not have any idea what enlightenment is.

Atomic–Consciousness

JAMES BATHURST

Atomic–Consciousness was a book published in 1892 by a man named James Bathurst. James Bathurst, the pseudonym for John Paltrey

(1846 – 1921), was born in Whimple, Devon England and was the eldest of six children. His mother was Mary Ann Hitchcock – a superstitious seamstress – and his father, John senior, worked for many years as a stonemason. Bathurst's book will be mentioned throughout these pages and extracts will be liberally taken from it. Although written in the later 19th century, Bathurst chronicles in depth many events that are highly pertinent to the work at hand.

Atomic–Consciousness, although mostly unknown today and of little note at the time of publication, was many decades ahead of its time. Even though much of the book is devoted to the minutia of the author's daily life and is filled with many outlandish personal theories hidden throughout this perplexing work are several ground–breaking scientific observations which anticipated quantum physics and the deeper understanding of human consciousness.

Atomic consciousness is not the same as cosmic consciousness. Atomic consciousness refers to the ability of all objects – both material and spiritual – to be able to think and perceive on some level as being distinct from the all-pervading infinite mind attributed to cosmic consciousness. While the conscious objects may be directed by the infinite mind, they do possess an independence which allows them to act individually when necessary.

Atomic-Consciousness originated in the mind of a man who had no formal education and labored throughout life among the working classes of England. He certainly never associated among the intellectuals of the day, spending most of his time working as a delivery man, railroad platelayer like his father once had done, and performing other assorted jobs that required enormous physical expenditure. Yet he somehow still squeezed out the time to write his opus. In this, he must be given great credit for resolve and persistence and overcoming tremendous literary, educational and financial obstacles.

Mr. Bathurst's acidic character acted as a potent impediment to a career as a philosopher and writer. He spent a great deal of his time complaining about his treatment and reviling humanity in general. At one point, he was diagnosed as insane due to his inability to conform to accepted "normal" behavior and he was confined to an asylum. However, it seems clear from portions of his writings that although he may have been troubled by debilitating psychological psychoses he possessed a truly original and gifted mind with extraordinary scientific and philosophical insight.

Atomic–Consciousness is vital to this current work because Mr. Bathurst applied a uniquely scientific explanation to spiritual concepts in a way which no one had done before. And since the publication of his book, only few have followed his path and echoed many of his theories. It may be extremely difficult for the modern reader to draw the comparison between Bathurst's description of the role that consciousness has on the shaping of the reality around us with the observations of great men such as Yogananda and Maharaj, but after careful interpretation in many instances they are saying the same things.

Bathurst's specific observations will be noted throughout. For now, his description of the concept of atomic–consciousness will suffice to demonstrate both his almost inscrutable writing style –which I have labored long hours over decrypting – and the depth of his thought process.

The following is written by James Bathurst and the italics are his:

"From long experience and observation, I am satisfied that there can be no other explanation than that *matter is conscious*.

"This theory will explain much of the phenomena surrounding life, and the laws which govern its various aspects.

"The original of matter, constitutes in itself a *polar* and *differentiating* force, from which has evolved the universe.

"*Mind*, being conceived as a particular condition of molecular structure, necessarily retains this property.

"Being therefore a part of conscious matter, it is an energy which, in the form of *will* or *ideal*, affects Atomic–consciousness, either in a *positive* or *negative* sense: so, we accordingly obtain corresponding results, whether such be acceptable or not, harmonious or antagonistic.

"Hence, *thought* is a *cause* which must produce *effects*; and whenever one *thinks*, Atomic–consciousness is influenced with a *force* corresponding to the *strength* or *power* of the mind from which it is generated."

AND

"Sensations produce *ideas* and *thoughts*, which create in themselves mental forces; and those being polar, the *effect* of such *cause* will correspond to the particular *state* or *form* mentally conceived.

"*Mind* and *Atomic–consciousness* are typical of two electrical substances positively or negatively charged.

"*Unlike* states attract and *Like repel* each other; and on this principle, those properties blend and are interactive. In virtue of *differentiation*, which characterizes the complicated organism, as well as the original atom, the stronger *repels* or *subjects* the weaker; hence the continuous war Between Mind and Atomic–consciousness, as shown herein.

"The union of atoms, for forming what we call *matter*, in all its shapes and phases, are probably results of this *subjection* – complexity following simplicity.

"Resolving that *mind* is *matter*, *thought* is *atomic*, or *molecular motion.*"

A similar "concept" can be found in the writings and teachings of the great spiritual leader Sri Nisargadatta Maharaj who almost echoed Bathurst's words in a talk given on November 21, 1980. "The basis and source of consciousness is in the material. ...The experience of consciousness experiencing itself." As can be seen, this is one of the ways in which scientific expressions and spirituality often cross paths and end up leading to the same conclusion.

There are numerous passages throughout Bathurst's book which may explore into quantum level ideas but which seem disjointed and bewildering, however, when isolated and read in combination with one another a meaning of profound significance emerges. This is like reading Bible verses from various portions of the Holy Book to capture the full impact that is imparted when studied together as one. Thus, putting together the psychic and scientific puzzle left by James Bathurst it is a very time consuming process since an attempt to view the whole picture has never been performed. And there are no concordances – reference books which show the locations of and relationships among Bible passages – for *Atomic–Consciousness*.

Another daunting feature about Bathurst's thinking process is that the concepts he professes must often be considered in reverse of how they are stated to fully understand them. In other words, the point that he is making is the often the opposite of what he is stating. Whether this is done by design – as a form of encryption – or by the accident of convoluted thinking, which twists into a figure eight to provide a circle, it is difficult to know. Almost as difficult as translating his words into understandable ideas. But it is worth the endeavor.

Infinite Presence as stated

WHAT DOES CONSCIOUS MEAN?

There are many definitions of what it means to be conscious. For the purposes of this book consciousness has the following meaning. The ability for any object that exists in any reality of any size – either material or spiritual in nature – to possess the innate power to communicate with or to accept communication from all other objects that exist, even from the highest Supreme Intelligence and to any degree of awareness.

Consciousness occurs on various levels of strength on a descending order, depending on viewpoint taken – being of a lesser strength and lesser awareness the farther removed it is in elemental form from the Primary Source, which can be described as God, Infinite Presence, or the First Mover Principle.

All consciousness is indestructible and exists on a quantifiable atomic and sub-atomic level and originates from the Source which retains its full infinite quantity throughout eternity. Such is how consciousness is defined in this book.

EVERYTHING IS CONSCIOUS

Everything in existence is conscious and related to the Infinite Presence (God), divine intelligence, First Mover Principle. This includes such items as television sets and sewing machines. This does not mean that the television set perceives you watching it or the sewing machine gets frustrated at your mistakes. What it means is that these two objects have the conscious ability to communicate with other conscious objectson an atomic level and can receive impulses from what will be described as the Infinite Presence because their consciousness is a minute facetof infinite consciousness.

Everything is conscious, including all material forms. How is this possible? Consider that human beings are conscious. Human beings are material forms. Yet, as material forms, we have consciousness. Why should not a rock or a tree or a golf club have consciousness of some type? Does the fact that they are material objects prevent them from having consciousness? We are material beings and are conscious. Why shouldn't other material objects have consciousness?

The type of consciousness being referred to is often of an electro–magnetic nature which acts as a communication medium between all other minds upon the planet which all possession a certain form of consciousness. Not all levels of consciousness are the same; some have less awareness of self than others. This type of consciousness is not to be confused with EGO which is a different matter entirely.

It is also not the same type of consciousness that is usually described as directing the so–called individual actions in a fully illusory world as professed by many of the spiritual leaders, such a Maharaj. The consciousness being described in this book is one that is ultimately directed from the Source with the intent of integrating communication among all levels of reality to maintain its control and balance over all things. We and the other objects are the channels through which this information flows.

There is proof that atoms possess consciousness. This was supplied by the double slit quantum physics experiment. Beams of photons were shot through a screen that had two side by side slits cut into it and, when not under direct observation, upon passing through the slits at the same time they appeared on a background target in a random chaotic disorder. This shocked researchers who had expected the photons to have simply formed two vertical lines on the background target because the light particles were provided only two side by side slits through which to pass. Yet the photons "chose" to assemble in scattered confusion onto the background target. This was their natural inclination.

To examine the problem more closely, an electron camera was set up to observe the double slits in the screen to determine what exactly was taking place. However, once this was done and the photons were ejected through the double slit screen again they fell into the expected double vertical line pattern as was originally anticipated. The researchers were even more startled.

What had happened was that the photons were consciously aware they were being observed and created the simple vertical line patterns that the researchers were expecting. Thought and control was demonstrated by photons on the atomic level! This conscious thought must have originated from the Infinite Presence which is part of all things.

Once again, we can turn to the teachings of Sri Nisargadatta Maharaj for a spiritual view on this topic. In a talk given on January 31, 1981 he had the following to say on this very subject: "The observer is also changing. What is being observed brings out a change in the observer, and unless that change is brought about in the observer, the observer cannot observe the object; therefore no one can ever get to the depth of spirituality."

A striking shared feature between quantum physics and the process of spirituality is the extreme importance of the observer in any situation.

The typeof consciousness that in so-called "lower" or elemental forms is not the same as in higher forms of existence. A toaster would not be expected to have the same level of consciousness as a frog and so on. Humans seem to have developed a type of consciousness that is more in tune with the Infinite Presence. Being aware of being aware.

Is that special element of human consciousness spiritual as revealed in the soul? That depends on how spiritual and soul are described. Cannot a rock be spiritual? According to animatism, it can. Not to be confused with animism, animatism is the so-called most "primitive" of belief systems which assumes that all things have a soul and consciousness. Primitive!

But when taking spirituality into account, since the spirit called God is clearly conscious and everything that exists has sprung from his being this would necessitate that all things share in this consciousness. Everything is one thing on all levels, though not equal levels, depending on the distance from the Ultimate Source.

From a quantum level viewpoint, Infinite Presence exists at the top of the wave function, extending to the bottom of the wave function and occupying all points in between and to the sides, controlling all actions that take place at any place and any time.

TYPES OF CONSCIOUSNESS

Infinite Presence is a force, a driving impetus that controls all things, all actions, all life, all potentialities, all futures and all pasts. It is pure consciousness itself and controls all realities by methods that cannot even be conceived by the human mind – yet. It is what ultimately connects all consciousness and uses various attributes to do so. It uses a delegation of power, just as God the Father does per the Christian faith. This will be examined shortly.

Is Infinite Presence God? It can be. It would be difficult to find any other description for such a force. But is it a personal God? Does it care? It is difficult to answer that. However, by studying the impacts Infinite Presence has on reality it could be determined that there is some form of consciousness to its working. Is it for good? Is it for evil? As noted, the best answers can be found in the effects it has on individuals and reality as a whole. Evil, however, is not one of its facets.

Infinite Presence does seem to function to assist us in our daily lives, no matter the occasion, even the simplest everyday event. Reality is constructed of the ordinary as well as the extraordinary. Relying on a religious comparison again, did not Jesus note how the Father regarded even the fall of the smallest sparrow? This implies interaction of Infinite Presence with every event that takes place on all levels of reality. All parts, fitting together. All consciousness in unison.

Any person can see the effects of Infinite Presence by examining his own past life, immediate and distant. Note examples of apparent serendipity. If you study any serendipitous occurrence deeply enough you may find a cause to it that was created by the effect. The effect preceding the cause. It is like saying God always was and always will be, beginning without end, created before existing. Can something exist before it began? Yes. Examples will be forthcoming.

There is a related action whereby consciousness overlaps and messages are passed between individuals and sometimes groups. It is like when you are attempting to contact another person on a telephone or I–phone and find that the person you are intending to contact is already on the phone attempting to contact you.

The following is an example of when mental messages cross in transport and is a form of an effect creating the cause. For example, a friend of mine found a recent photograph of himself in which he noted how he thought he looked like the actor Charlie Sheen. At the same time, his daughter had already sent him a photograph by email of Charlie Sheen who was at a baseball game and noted how her father looked like him. Neither of them had any discussion or contact about Charlie Sheen beforehand. Was this just "coincidence"? It would be easy to answer, yes. Except this type of happening takes place on far too regular of a basis for it to be coincidence.

What is the purpose of this second form of transmission of consciousness? It primarily has a prophetic quality, but occurs on numerous levels, like the one just highlighted. All prophecies need not be world changing. But everyday types of premonitions occur often. Continuously. They are meant to help us in our lives.

Infinite Presence is in control of these occurrences. One basic purpose is to maintain the flow of all possible realities, per the will of Infinite Presence. We can only hope to understand a concept that controls the existence of all realities by being able to observe the effects and know that there is a reason and purpose behind them. Infinite Presence causes all things to happen and thereby maintains its own existence as well as ours on all levels.

Thoughts are contained within consciousness. Therefore, since consciousness can affect material things, consciousness must also have a material aspect. And as such, so must thoughts.

Thoughts can be seen to possess both a negative and positive factor just like magnets have a positive and negative pole. It is in the way that thoughts attach to other thoughts or other material things, that ideas are created or actions are initiated.

In this way, certain thoughts create certain ideas that are beneficial to a person while other thoughts create certain ideas that are not beneficial to that person. This is a very complex process which is covered in depth in the section about the Neutral Mind. However, one of the important points to be considered at this juncture is that neutrality of thought is what provides the best alternative in any situation. It is even to be preferred over what might sometimes be termed positive thought. The reason for this is that positive thought can be diminished by contact with negative thought. This occurs because, as in the magnet, unlike poles tend to attract. Positive thought can attract or be attracted by negative thought with the consequence being a persistent uneasy lack of balance between them. Therefore, neutrality of thought is always to be preferred.

There will be much more on this later, however, before leaving this topic an example will be given to demonstrate how the power of thought when developed into ideas can be transmitted and spread among groups of people, occasionally causing havoc, based solely on the power of "mental" energy. It cannot be said to be non–physical because this type of energy is sub–atomically activated.

From the *Daily Press*: "A singular nervous disease has recently been affecting, in a terrible manner, German children in day schools. The first commencement arose through some scholars, in one of the establishments spreading the rumor that 'ghosts' haunted the place. Rapidly the children attending *other* (authors italics) schools were affected, and the rumor turned to a general belief. Extending from school to school, it became a panic, large numbers being ill from fright. One would imagine that a 'ghost' was waiting to seize him; another that she was really in the clutches of that monster." Was this a matter of pure thought that created the existence of ghosts throughout a German school district? How did these "imaginary" ghosts appear as widely as they did without any communication among the sites?

UNIFIED–CONSCIOUSNESS

In circumstances when one form of consciousness contacts another, the Infinite Presence uses what I term Unified–consciousness to perform the union, acting as the go–between among all other levels of consciousness to transmit information, even of the most common nature.

Why would Infinite Presence (or God) use an intermediary to perform this work? Certainly, the ultimate intelligence could simply complete the task itself. True, but Infinite Presence may also work on the basic principles of electronics. If a source of power is downloaded to a greatly inferior receptor of power, the inferior receptor (human or another object) might be destroyed or seriously damaged. Not every contact with the Ultimate Power need be a miracle, some can be a simple cosmic jolt of hope. Thus, Infinite Presence will rely on a more limited display of energy to conduct whatever flow of data that is required for the situation.

In religious terms: God is known for using his angels as messengers and to carry out other instructions. Angels as electro-magnetic conscious energy will be fully examined in Part 2.

Certain unique terms will be used throughout this work to describe phenomena that are described differently in other source material on this topic of belief and science. The two major forces are Infinite Presence, which implies a conscious entity which is omnipotent and benevolent

(Godlike) and Unified–consciousness (the Holy Spirit?) which is an attribute of Infinite Presence which is used to disseminate instructions to and among the lesser levels of consciousness. These are at this point arbitrary terms created for this report. Many other similar terms will appear throughout the following pages. If you are of a non-believing nature, you do not have to accept the idea that Infinite Presence is conscious.

Ultimately, everything is one thing and refers to the Infinite Presence. This is scientifically verified by the reality of quantum entanglement whereby particles or groups of particles no matter where they may exist at any location in any universe cannot be independently identified as being different from one another. If one of the entangled particles is stimulated, its counterpart in another location will react the same way.

Evidence of Infinite Presence

One of the less important actions of Infinite Presence is to produce the realization of desires made to it by an individual if it is in accord with the ongoing construct of reality. This can involve either large matters or less important matters, generally of the last type. Is not God in tune with even the most basic aspect of creation? As already noted, in religious terms, Jesus stated that the Father was so precise in this as to even numbering the hairs on a person's head.

Infinite Presence is interacted with by the thought process of the individual. It does not have to be in the form of a prayer or of stylized ritual. A simple thought in the mind is all that is required. But it is important to understand that it is not the person who creates the results but it is Infinite Presence.

NON-CAUSALITY

The following is a personal experience which caused me – after a vast number of previous similar occurrences – to finally accept this concept as a fact. Non-causality implies that a cause is produced by an effect, rather than the traditional scenario where the effect is the result of a cause. I was awestruck when I realized the fulfillment of a wish which was granted simply by mentally asking. Or, in biblical terms, "Seek and ye shall find." I did not have to state it aloud. But in addition to this it will be shown to have been a matter of an effect preceding a cause. In biblical terms this might be called prophecy, or in layman's terms, precognition.

In the month of June 2016, I read the book *Atomic–Consciousness* by James Bathurst, which as noted provided a major premise for the work you are now reading. I was so overwhelmed by the proposals in Bathurst's theory that I sought to learn all I could about the author. Nothing was available. Being an adept researcher, I was unused to this type of stark dead end. But I simply could find nothing. It seemed that no one had any data concerning James Bathurst and that he must remain to me an enigmatic figure from the second half of the 19th century. Yet, I still greatly longed for any information about him.

At the time, I could not have known that another researcher – named Daniel Wilson – an expert in acoustics, and in tracking down hard-to-locate authors of the past, had also discovered the existence of *Atomic-Consciousness* well before I had. Also, that he had succeeded in finding information about the author. He wrote an authoritative article about James Bathurst/John Palfrey and in 2009 first proposed an article about atomic-consciousness to ForteanTimes Magazine. No action was taken by the magazine. Mr. Wilson submitted the article again in October of 2015. The piece remained unused.

Then, finally, almost magically FoteanTimes found room for the article and published it in its June 2016 edition. The same month I had finished reading *Atomic-Consciousness.* The same month in which I was fervently seeking information about the author.

It was only accidentally that I even knew Mr. Wilson's ForteanTimes article existed. By what would otherwise appear to be random searching of the internet for unrelated material I happened upon a blog written by Mr. Wilson, called *"Miraculous Agitations"* in which he addressed the background of James Bathurst in detail. I was thunderstruck when I found this! How could this possibly be a coincidence! How could it possibly be serendipity? Basic probability factors would rule this out. My discovery of this article now was an occurrence of duality of action which had to have been directed to take place by what I have previously identified as Unified-consciousness. Mr. Wilson and I were in 2016 both professionally considering James Bathurst's *Atomic-Consciousness* at the same time without each other's knowledge.

Unified–consciousness links human consciousness to all other levels of consciousness and is ultimately directed by Infinite Presence which oversees everything. In religious terms, Unified–consciousness would assume the personality of the Holy Ghost, or Holy Spirit, which, under the direction of Infinite Presence, or God, is the transmitter of knowledge and other types of information among all the other levels of consciousness that exists. Angels would perform a similar task in a similar though not identical way. In ancient Roman religion, Unified–consciousness would be the god, Mercury, the messenger of the deities.

But there's even more. A great deal more. The *Atomic–Consciousness* example provides three specific qualities attributable to Infinite Presence: fulfillment of a desire or wish, duality of action, and a case where an effect produced the cause. The fulfillment of desire occurred when I found the article written by Mr. Wilson. The duality of occurrence took place in the fact that both Mr. Wilson and I were professionally dealing with the concept of *Atomic–Consciousness* and its author, at approximately the same time without foreknowledge. And my discovery of the book *Atomic–Consciousness* and my stated desire to learn about its author was an effect that produced a cause.

How was the latter an effect produced by a cause? Mr. Wilson's article was written years prior to my even reading the book *Atomic–Consciousness* and as such it represents an answer to a wish that would not be made until the future. This to me seems quite profound and impossible to ignore.

But as significantly is the fact that the publication date of the article in the ForteanTimes was delayed by *__7 years__* after its initial creation by

Mr. Wilson. The publication date – June of 2016 – coincided exactly with the dates I read *Atomic-Consciousness* and my stated desires to acquire the much lacking information about its author! My wish from the future fulfilled by an event from the past? This concept is made much more secure by the 7 year delay in the ForteanTimes publication of the article. Just coincidence that the piece did not appear in the magazine until 2016, or a series of interconnected consciously directed events, including a summons from the future?

Such outcomes were quite clearly defined in *Atomic–Consciousness*. This is the personal account of such an event by James Bathurst. "If any matter engaged my thoughts to–day, which in all probability could only be satisfactorily answered or explained by a public newspaper, journal, or book; before many days, there would appear in print, and in such a way which would secure my attention, the very subject."

Continuing: "Quite frequently too it is observed, in a manner which proves that it must have been produced *before* the desire was generated, thus proving Atomic–consciousness sensibility." Thus, the written material he wished to see had been created prior to his request to fill his request. Of course, this could be just a coincidence, but these types of so–called coincidences happen too frequently to be coincidences per the laws of probability.

WISH FULFILLMENT

Infinite Presence also operates under far less consequential conditions as well. In fact, under conditions as mundane as the selection of a specific movie to fulfill one's request. There was a period in my life when I became highly interested in the life of an actor named John Chandler. He exclusively played in minor roles on television primarily during the 1960's and the 1970's. Upon learning that he had once had a starring and lead role in a movie titled "Mad Dog Col" it became a very strong desire of mine to see this film. But I couldn't find it listed for view anywhere or the video for sale anywhere.

Then, as if in answer to my request, or prayer, at this precise time the movie "Mad Dog Col" appeared on the TCM (Turner Classic Movies) station. It had never appeared here before, and to my knowledge it has never been re-shown. But at the exact moment I was hoping to see this rare film it appeared on TCM. Naturally, this could have been a coincidence. But what are the odds of it appearing at the exact time I was hoping to see it and none other? And when considering that this same type of scenario has played out for me on numerous occasions it is difficult to assign this to the category of coincidence.

How to account for it? Some magical power I possess? Or is it an answer made by Infinite Presence to a request I made? Why such a trivial matter? One might just as well ask, why not? Daily life is constructed upon trivial matters. Remember the words of Jesus. And maybe the "trivial" in connection with other events that we know nothing about isn't truly so trivial. Maybe even the least significant appearing events have great significance when placed within the entire framework of reality.

I am certainly not the only one to have this experience. Probably many who are reading this have had the same experience many times. As well as friends and relatives. And many people have had wishes fulfilled without even being aware of it. James Bathurst was clearly aware when his desires had been met and he carefully chronicled most of them.

One of them concerned his great desire that a University lecture course that was being promoted by Oxford and Cambridge Universities would dispatch lecturers to his town of Bath. These courses were taught by professors who travelled throughout England, offering free lectures to the working poor men and women of the community. And much to his delight, this series did come to his town shortly after his mental request. Was it in answer to his request to the Infinite Presence? Or was it, again, merely coincidence?

There are many other methods by which we are contacted or by which we contact the Infinite Presence. Sometimes we request a wish to be fulfilled. But what happens when something that a person longs for or desires to happen is not in accord with ongoing reality? A process called Negation occurs. The wish or desire is denied or else in some way thwarted and kept from full realization. In religious terms: God's will. But God didn't do it. It was simply something that didn't coincide with the ongoing event of the multi–verse taking place because if it did it would dislodge certain other actions out of balance. And if a person were to fixate and obsess upon this frustrated desire he would then be the creator of his own misery.

LAW OF ATTRACTION

James Bathurst further observes in his 1892 book: "When we have once received a desirable thing, the knowledge or consciousness of having so received it is a mental subjection (attraction) of that fact. Thus, it repeats due to Atomic–consciousness (will of God?) One benefit or action is the cause of others (similar events) (law of attraction?). But, once happy or content, differentiation (Negation) occurs and one will suffer. Those who fail to receive or get advantages and are conscious of these failures cause these to persist. Memory of the past in relation to one's success or failures produces corresponding repetition."

As commonly understood, the law of attraction involves an individual requesting the universe to provide for him the desires he wishes and that this becomes more likely and possible when this person develops and maintains a positive attitude in this regard. His will acts upon the universe in a complimentary way, thus attracting to him the benefits which he is seeking. It is a spiritual form of supplication of the universe which will then hopefully respond reciprocally.

However, there is also a scientific course toward that same end which can be followed which uses a different technique but which bears the same positive results. Even though positive results are effectuated in both methods they are viewed as being obtained through a slightly different process.

The scientific process is basic psychology, but it is decidedly applied psychology complimented by biological effects. It is more than a matter of becoming psychically in tuned with the universe on a positive level. It is a technique of following a specified set of guidelines which leads to a goal, even though the outcome will have an aura of the spiritual about it.

The psychological method begins with creative visualization. The person envisions what positive event he wishes to happen and sets this as a goal. Having a specific goal is vital. From that point onward reaching this goal is a matter of directed concentration on it and how it will be attained. In addition, the person must focus on having attained this goal and what that experience is like. So far, this is rather basic.

What is significant about this approach, however, is that it has been proven by scientific measurement to be effective. It affects both the person who is attempting to attract success but it also affects those around him which helps him acquire success. Spiritually, it could be likened to another person being enwrapped within an aura of positive reflection. Biologically, it can be termed as a sharing of positively charged energy which is being transmitted by the consciousness of the individual seeking positive change. And it can be measured.

This sharing of positive energy causes other people to react differently to the one seeking change. This results in a variety of actions that then occur which otherwise may not have taken place. It also causes other people to be more positively attracted to the seeking individual which makes it much more likely that the goal he is striving for will be successful.

And of course, negative energy acts in just the opposite way.

NEGATION

Negation is a highly significant process since it will be the cause of most of a person's sadness. Desire of that which is not attainable because it does not coincide with the ongoing construct of the multi–verse. It is like a contradiction. It is like a bruise that keeps being struck repeatedly which not only does not heal but worsens. It is like anti–matter colliding with positive matter. In a case where seemingly negative things happen to a person it might be best to remember Krishnamurti's attitude of "Not minding what happens" because what ultimately results are often to that person's best interest.

It is when a prayer is not answered because if answering it a negative set of actions would be produced which were not intended by Infinite Presence to occur. What follows is a perfect example of Negation. It is an event again taken from the experiences of James Bathurst, whose life is highlighted not only because of the care with which he recorded the events but because he understood and could explain what was truly happening to him.

This is taken again from the book *Atomic–Consciousness* and is written in Mr. Bathurst's own words. "For some time I have been interested in newspaper reports, noticed frequently, regarding an occult science propagated by Madame Blavatsky, called *Theosophy*. The effects of this woman's presence, in a room, were such as to convince even the skeptical, that she possessed some extraordinary and mysterious powers, and that apparently, there were beings of another world, with whom she held communication. On becoming acquainted with these facts, I earnestly desired to be a witness, and, if possible, ascertain whether the effects were deceptive. Having no friends through which an interview could be effected, my hopes were ever met with disappointment."

38

Then occurs what would fall into the realm of the above noted form of wish fulfillment. Mr. Bathurst's desires were fulfilled, seemingly by chance, although not truly so. Unified–consciousness was directed to make the following action take place.

Continuing with Mr. Bathurst's account: "Last month while performing some work for Mr. G—— of a temporary nature, the gentleman became, by some means (Unified–consciousness?) much interested, but from what cause I cannot say; possibly through my pensive countenance and reserved disposition. Among other things, he induced me to mention *Theosophy* and expressed his willingness to defray my expenses."

Thus, the travel costs for Mr. Bathurst to attend the meeting with Madame Blavatsky were covered.

Unfortunately, the action of <u>Negation</u> suddenly erased Bathurst's hopes. The meeting with Madame Blavatsky was to have occurred on May 11th. She passed away on May 9th, two days before.

However, Bathurst has an additional explanation as to why Negation takes place even if a person could reverse its affects."It frequently happens with those who have strong conceptive or idealistic powers that should one regret or feel remorseful for any unfortunate occurrence and at the same time imagine the reverse taking place this mental state is often sufficient to cause events to occur precisely in accordance with one's imagination. BUT WHY DOES THIS NOT OCCUR ALL THE TIME? If it did, wouldn't this be a perfect world? One would not have cause to be angry, or fight or get involved in any unhappy situation if every wish was fulfilled."

NEGATIVE REALITIES

The forces of consciousness can assume a negative – some say evil – character if they are altered by a person. This negative vibration can be transmitted to and become part of any animate or inanimate object. It can also be retained as negative memory, or energy, in the environment. Cite the following newspaper accounts.

"Three years ago Lieutenant Roper, of the Royal Artillery, was shot at Chatham with another officer's revolver, but by whom the deed was done, nobody could ever tell. A few weeks ago, a gentleman in India, who afterwards got the instrument, shot himself with it."

Often, as noted, negative thought energy can also become attached to places as in the next two events. "A man was killed at a spot on the Great Western Railway, near Bathampton village, a fortnight ago. Today, another was killed at the same place."

In 1664 a boat crossing the Menai Straits was lost and one man drowned, Hugh Miller. In 1785 another boat was lost in the same place and one man drowned, named Hugh Miller. In 1820 a third accident of the same type occurred in the same place, and one man died, named Hugh Miller. *Daily Telegraph*, Jan 12, 1888. No, these were not typos or misprints. Over a span of three centuries three different and unrelated people named Hugh Miller were killed in boating mishaps at the same location in the Menai Straits!

The following tragedies occurred at the same location or close by. "The Reverend J.H. Davis, a Vicar of Bishop Burton, has committed suicide. A few years ago, a gamekeeper in his service, named Jex, was murdered, and several others injured. Another case occurred, in which the Vicar's sister left the town a bride, and was shortly after found drowned. Another minister of the same place went mad. A lady, who was wife of another Vicar, severed the artery of her arm, and bled to death." *London echo.*

It should here be noted that one of the explanations for the existence of ghosts is that the memory of their physical lives left a material impression upon the environment in which they had once existed and that this can still be experienced by anyone who encounters this location. Sometimes, this effect on the environment exists as a potential for animation and does not become active until there is someone to observe the phenomena. This is very like quantum physics in which a reality can only be said to exist once it has been observed.

MATERIAL PRECOGNITION

In some cases, when destructive events occur it may either be an example of Negation, or possibly an effect preceding a cause, or perhaps even a prophecy sent by a higher intelligence, Infinite Presence. There is a famous historical instance of which there is physical evidence that something of a prophetic warning occurred but seemed neither to have been noted nor understood nor acted upon.

It involves the ocean liner the Titanic. As most people know, the Titanic was sunk when it slammed into a massive iceberg. Some people do not know that the Titanic hit this iceberg because the ship was traveling at an unusually high rate of speed at the time for that part of the ocean.

And fewer people probably know that this speed was so high because one of the coal storage bins had caught aflame and to reduce the risk of explosion the coal was hurriedly shoveled into the boilers which added to the speed of the ship. And most everyone knows that the Titanic then struck an iceberg and sank.

What very few people also do not know is that the exact spot on the ship which was punctured by the iceberg had already suffered damage before the ship was even launched, leaving it even more vulnerable to further damage. In previously unknown photographs (released in 2016) taken of the great liner before its departure, there was discovered a sizeable discoloration on the outside of the hull at the exact location where the iceberg was to later strike. The supposition is that the fire in the coal storage bin had already begun smoldering before the ship even left port. Another explanation could be that this is an example of an effect occurring before a cause. The initial damage from the iceberg strike had already appeared, maybe as a warning from a benevolent superseding intelligence.

The tragic sinking of the Titanic was predicted in a novella that was written in 1898 called "*Futility.*" The author was Morgan Robertson. The Titanic sank in 1914 but many of the events described in the 1898 book were too accurate to be accidental.

The similarities between incidents are the following. In the book the ship was called the Titan and was the largest ever constructed until that time. Both ships were almost the same size with the Titanic being a mere 25 meters longer. Both were supposed to be unsinkable. Both were sunk by striking an iceberg in mid–April. Both could travel at a speed of over 20 knots, unheard of in that day. Neither ship had enough lifeboats. The iceberg hit both ships on the starboard side. Each sunk at approximately 400 nautical miles from Newfoundland. Both ships were of the triple screw propeller type. Both ships were travelling at a far faster speed than deemed safe for those waters at the time each collided with the iceberg.

Coincidence? In religious or spiritual terms, this would classify as a premonition sent by God or divine intelligence or Infinite Presence, using all–seeing powers and dominion over time.

But when using Infinite Presence as the originator of the transmission of information, it is a matter of data from a future event being sent to the past – by way of Unified–consciousness – and to the mind of the author. The purpose was to provide a warning of the upcoming disaster. And this would represent a case of effect preceding the cause.

There exists at least one scientific explanation for this godlike action by Infinite Presence. The existence of tachyons. Tachyons are particles that travel backwards in time by being able to travel faster than the speed of light. Impossible to travel that fast? According to Einstein, maybe. But, Einstein is not God, or Infinite Presence. The Creator is not bound by any rules because he makes the rules.

One of the primary arguments against tachyons having the ability to travel faster than the speed of light is that it violates the law of causality which says that the cause must precede the effect. (It will be demonstrated in this book that this is not true). However, those who believe in this law – which is a law of classical physics – seem to fail to understand that they are thereby accepting a paradox as their argument against another so–called paradox. The speed of light is a paradox when it is shown that its speed, which is supposedly fixed and the fastest velocity attainable, can be superseded by certain particles in certain instances.

One specific paradox concerns electro–magnetic brain activity. The paradox alluded to was demonstrated by Gary Weber in experiments involving the brain's ability to foresee a future event. In this test, individuals were asked to make a motion of their hands in response to a stimulus that was to be given. The voluntary hand movements were made in response to this stimulus but the movements occurred before the stimulus was even produced or activated. In other words, the thought process was of such speed that it anticipated the stimulus and reacted appropriately to it even before the stimulus occurred. This implies that the effect took place before the cause. Did not this violate the law of causality, even on the macro level?

And, as a final observation on the matter, are the writings of James Bathurst from *Atomic–Consciousness*. "The sensibility of atomic–consciousness is so great that the *effect* of some mental state yet to take place at a future time, is such that *re–action* or results arise, *before or antecedent* to the originating cause, the interval and frequency depending on the organic state or mental susceptibility." (1892).

Contacting Infinite Presence

THE NEUTRAL MIND

The neutral Mind is one of the most important and powerful concepts that will be presented in this work. It allows a person control of his thoughts and environment, and it will produce desired effects in the future. It is a simple matter of allowing the mind to assume a passive state where it allows the consciousness around it to perform its function and to create the effects that are destined to be created.

James Bathurst: "For a person to be successful, at no time must there be any given sign or indication of anxiety or care – in fact, one must be indifferent and heedless whether a given thing takes place or not. For the hope or the want will produce Negation."

This is not an uncommon belief, and is one of the major concepts proposed by such great men as Yogananda. However, what is the actual, physical mechanism involved in this process? Is there one? How to explain in strictly scientific terms what the Neutral Mind implies? While the concept has been earlier considered, it might be wise to revisit the atomic level nature of the Neutral Mind at this point.

Thoughts, composed of atomic structure, carry either a negative or a positive charge. Consciousness consists of thoughts and is infinite and indestructible even though existing on various levels. As already noted, each thought is like a magnet, having both a negative side and a positive side. And these can attract either negative ideas or positive ideas, both types of which are also of an atomic nature. But, like a magnet, the negative idea repels the negative and the positive idea repels the positive. While a person may be seeking pleasure and happiness with a positive thought it is naturally attracted to the negative counterpart of this condition and, like a magnet, is captured by the negative and thus they combine into an uneasy and often combative pair. (Is this the origin of ego?)

However, if a thought can be made to assume a neutral charge it will then attract only thoughts of a positive aspect because the nature of a positive thought is much more compatible than a negative thought. Once a state of neutrality is realized by the mind an overall character of peace and harmony will result and happiness can be more easily obtained.

How can thoughts be changed into a neutral state? Primarily through meditation. Meditation has proved to be able to alter the physical workings of the brain. That is why so much emphasis is placed in meditation by all practitioners of any viable spiritual belief. It works!

Not everyone may know the atomic level at which meditation alters thought function, but the effects are what is important. And this is not to discount any spiritual study and preparation that is part of the process. Will a person always be happy? No. Will all her goals and wishes be realized? No. At least not until total control will be attained, which is a very difficult level of consciousness to reach.

Is there any real physical proof of any of this? Yes. It is revealed in the fact of quantum level entanglement. When an atom is split, and separated in space each half can be affected by a stimulus applied to the other half. This shows that atomic structures can be influenced by outside stimuli.

Infinite Presence is best influenced by contact with a consciousness that has attained a Neutral Mind. Therefore, that is when the best benefits can be expected to result in daily life by being in accord with Infinite Presence.

As already noted, it is possible to contact this all-powerful force. Since it is of a mental nature, the most efficient manner of contact is through thought. One of the best ways to do so is through meditation. My preferred form of meditation is Transcendental Deep Meditation as taught by the Maharishi Mahesh Yogi. It involves purging the control of thought by use of an individually selected mantra. The purpose is to escape the confines of thought and emerge into the world of pure consciousness.

But simply placing the mind in a hypnotic type of mode can be quite effectual. One form of hypnotic state canbe as easily reached by sitting quietly in a chair and listening to tranquil music in the background. This can allow the mind to drift to deeper layers of consciousness and potentially tap into Infinite Presence.

A personal experience is used as an example to illustrate how sublime yet logical this technique can function without true effort. The only effort required is lightly directed thought. And then, sometimes strange and mystifying phenomena will naturally occur.

From personal experience. Seated comfortably in a cozy chair in a darkened room with instrumental music in the background, my mind was drifting without any apparent purpose. Then, for some reason, thoughts focused upon a problem that needed a solution: what was the last name of one of the most famous early Welsh actors who made many movies in America in the 1930's and 1940's. His name refused to arise, even though

I knew him well from the television screen and it was a relatively common surname. Also, his last name was the same as one of the most famous of Welsh football players in the early 2010's. But no matter how deeply concentration was placed on this seemingly unimportant problem the identification of the surname of either man remained elusive.

At this point, Infinite Presence sent Unified–consciousness to assist. And keep in mind, even though the matter seems trivial, this form of assistance occurs frequently and should be recognized. In times of true need, one should be alert for this type of subconscious help. The instance currently being described might be used as a type of practice session.

While I still struggled to capture that Welsh surname a specific melody appeared in the background in the ongoing music. The name of the song was also unknown. But there was one thing distinctly unusual about it. Among the instruments being played were tubular bells. Tubular bells are a specific type of chiming instrument that were first created by Mike Oldfield for his classic LP, *Tubular Bells*.

The Oldfield LP was originally made in 1973 and it is doubtful that in 2016 many people under the age of 45 would recognize the sound of tubular bells. But I am a Mike Oldfield fan and did recognize the sound. And this provided the hint that was needed to solve the mysterious surname problem. The word bell was immediately transformed by the thoughts into the name of Bellamy. The actor was Ralph Bellamy. The football player is Greg Bellamy. And Unified–consciousness had fulfilled its mission,

Meaningless? Trite? Wishful thinking? Absurd even? But when all the factors which led to the solution of the elusive surname are considered their interconnection makes perfect sense. Specifically directed at a specific individual's understanding of the clues. And this type of contact with the underlying Infinite Presence happens regularly. Again, while this may seem too inconsequential to be of any importance – too trivial – none of us can truly judge what is and is not of extreme importance in the vast matrix of this and all other alternate realities which will shortly be considered.

But there is another totally opposite way that the answer to the elusive name could also have been arrived at through the operation of Infinite Presence. And it is the one most people are told to use when trying to remember something that is just out of reach. Forget about it. Put it aside. And when the mind is ready the answer will appear. It is another example of Neutral Mind concept which is such a powerful technique of mental silence. When control is surrendered, a solution to a problem will often be forthcoming. Just have faith that it will.

In fact, that is precisely when the tubular bell answer was given. When the mind was set at rest and became resigned to the idea that the solution would appear when it was meant to appear, the solution did arise. That was when Infinite Presence contacted the conscious element that was in control of setting the soundtracks to be played over the television to air the song which contained the tubular bells.

Once again we are faced with standard religious doctrines. Surrender the will to the greater power and have faith that an answer or help will be supplied. This, too, is highlighted repeatedly in the work by Bathurst. I consider it to be one of the most important of observations that were made in his book. It basically states, that the mind must be un–

encumbered from overly passionate wants or desires or fears to obtain that which is most sought. A state of expectant indifference. The following is Bathurst's description.

The law of Neutral Mind is important enough to be repeated. "For one to be successful at no time must there be any given sign or indication of anxiety or care – in fact, one must be indifferent and heedless whether a given thing takes place or not. For the hope or the want will produce differentiation. (Negation.)"

He then goes on to describe how if a person did possess the power to alter reality by merely wishing for perfect harmony, it could not ever be accomplished. It would mean an end to strife or misery. Next, is the second part of his answer to this question.

"The generated force of the mind is like that power from a battery. It follows the path of least resistance. Whatever one is indifferent to or heedless of, will, provided such indifference or carelessness be subjective, react through the differentiating law (Negation) to one's interest and benefit simply because one's will exercises positive influence and is at the same time fearless of converse results. Atomic–consciousness succeeds through individual doubt, fear, or timidity of victimizing the ego instead of the ego subordinating the Atomic–consciousness."

ELECTRO–MAGNETIC CONTACT

Electro-magnetic contact is very like the one just described. But its manner of working is even more direct and measurable. First, a question to the reader. Have you ever, while watching television, or while listening to the radio, or while being in contact with another form of electro-magnetically operated device, had the experience of simultaneously thinking the same words that had just been transmitted over the device to which you were listening or watching? This is a common experience for me. And the overlapping of thought is at the precise moment– simultaneously. For example, when someone on the television says the words, "I thought I'd go," you were thinking those exact words at the exact same moment but often in an unrelated context, "I thought I'd go." Are our minds tuned to certain vibratory frequencies with transmission reaching them from all points across the universe?

Some people try to explain this phenomenon the same way they explain Déjà vu. They claim that in fact the mind had already heard the words that were spoken and simply replayed them a microsecond later, fooling itself into thinking it had heard them simultaneously. Just like Déjà vu is supposed to be the remembrance of a thought that hadbeen just experienced a microsecond before. Déjà vu is another matter which will be discussed at even greater length later and will be properly identified as much more than the mere memory of a just experienced thought.

What this form of electro-magnetic communication involves is the direct contact between the radio waves being transmitted from the electronic device and to the brain of the recipient. In other words, it is a matter of the brain intercepting the electro-magnetic waves as they were either being directed toward the inanimate receiving devices themselves from their studios or from the devices themselves after they had received the transmission.

While a more metaphysical explanation for this occurrence has yet to be discovered – for example as representing some form of telepathy between the person transmitting the sound electronically from the transmitting studio to the person receiving it – this form of communication clearly demonstrates the ability to transfer data from an inanimate object to a thinking human being. What remains to be determined is whether the transmission from the inanimate device is a conscious effort from that device to establish communication with the intended recipient. Also, if this type of communication may in fact be a form of mechanical clairvoyance.

Another explanation is radionics. One aspect of radionics is theoretically to provide a form of direct thought communication by sound waves. By this process, thought patterns can be scientifically configured into electronic harmonic waves. It is a process that was developed in the early 20th century by Delawarr Laboratories. For example, if a person concentrated upon the thought, "I would like to see my friend Wilma tonight," it could be transformed into an inharmonic frequency tone cluster which could be transmitted to a receiver. The receiver, of course, could be the human brain, or mind if you wish.

What if Wilma captured and interpreted this thought? "I would like to see my friend Wilma tonight." Might it not then seem as if a premonition had been realized should she appear that same evening at the friend's house who had transmitted that thought?

Delawarr labs produced many thought wave messages. Curiously, from around the 1960's, one of the wave transmissions was the mental word "resentment" and was contained in one of the following frequency groups: 8Hz, 59Hz, 80Hz, 130Hz and 350Hz. There were other sets of frequencies, used as well.

There will be more to follow on this topic when the subject of premonitions and harmonics from the future are considered.

DIRECT CONTACT – MEDITATION

At the human level of consciousness, we can make direct contact with the Eternal, God, the Infinite Presence or that which is. This is primarily a state where the consciousness is directed on a path to discover the source or essence of all being, or non-being, using a mantra or other form of mental vehicle. There are various types of meditation techniques, but they basically are all aimed in the same direction – inner enlightenment.

Wondrous manifestations can result from practicing meditation, as already mentioned, and the results usually become even more profound in relation to the amount of time devoted to its performance. It is well beyond the scope of this work to provide information on the specific methods that are applied in meditation, thus it is highly suggested that if interested a person should seek out the type of practice that best suits him and his lifestyle. This author is a practitioner of Transcendental Deep Meditation as originally taught by the Maharishi Mahesh Yogi and can verify that its effects can lead one to places beyond thought control.

Again, most forms of meditation are valuable and will provide the truest and most direct path to that which is.

Recurrence and spiritual renewal

RE-ENACTMENT

This section is critically important in that it shows how personal life events are repeated and certain negative results can be avoided in the future by studying the past.

Are there important events in your life that seem to repeat themselves? Maybe not in the exact same way, but similar enough to make you wonder why they are so similar? Such events may have happened or may be happening but you may not have noticed. Pay attention, please, because it is possible that you are being given an opportunity to change situations for the better. Infinite Presence may be trying to help you in ways you are not aware of, but of which you can be. Observe closely the life events taking place around you and note the specific episodes within them. You may be surprised. They aren't accidents. And by changing them as they occur based on past occurrences you may be able to secure a better outcome now.

In this context, a life **event** is contained in a series of daily **episodes** which progress in a regulated order. The episodes are segmented actions which make up the whole event, which in this case means a life event that may take weeks or years to play out. A life event may be something like the slow dissolution of a love relationship or the loss of a business. Obviously, it is only the negative events that a person would seek to change and keep from recurring. The positive events one would wish to understand and repeat.

One of the primary features of Infinite Presence is to allow for the replay of life events of all types that provide the potential for positive change even from – or especially from – a negative life event. The outcome of each replay can be altered at any step along the way if the people involved understand the precautions that are being given as the process progresses. These warnings are being given by direction of Infinite Presence to lower level conscious objects to allow for change.

The life events will be of a major type – loss of love, finding a new career or changing one, etc. – but ultimately in the repetition phase of the initial event–the cause will be produced by the effect since it will initiate the event's pre–determined repetition. Sound confusing? It is. But it can be understood with a little bit of explanation.

An example will help clarify. Like James Bathurst in *Atomic–Consciousness* I will use personally verified examples to help prove the point. The event to be considered is a man's lover of many years leaving him. This was followed by a repetition of the same type of event with a second lover in almost the same way. This was related directly to me by a friend, Ed, and I also witnessed the event both times it happened. Unfortunately, this was many years ago and at the time I did not know enough about this specific subject to have helped him.

As noted, this type of event occurred at least twice to Ed. He had lived on separate occasions for about 10year periods each with the two women involved, Sue and Joan. In each case, both women in the relationship became overwhelmed by the powerful promises of the man who would replace the original, Ed, and in both instances they left to begin what they hoped would be greatly improved lives with unrealistic expectations.

This life event is a prime example of recurrence because the step by step episodes in each event was almost identical in nature. Due to my friend's lack of insight and his inability to react to the precautions he was being given, he did not change his behavior and thus the second event of lost love with Jane recurred almost exactly as it had in the initial event with Sue.

Many of the precautions Ed was given were of the most commonplace type. One of the repetitious episodes that took place during BOTH events would likely not even be noted by most people. But Ed is meticulous about observing such things. Especially things that seldom happen to him; he notes them when they do. Unfortunately, he didn't understand their real significance until too late.

One episode that occurred during the loss of his first lover, Sue, was when my friend had caused her to be awakened from a peaceful slumber by the sound of loud music which Ed had been listening to through headphones. He didn't realize that the headphones had stopped working and that the true sound was blasting through the loud speakers. This awoke his soon-to-be ex-lover, Sue.

During the loss of his second lover, Joan, the same situational episode took place, only on this occasion Ed had been listening to loud music through headphones attached to a television. But the result was the same, his then lover Joan was abruptly awakened when the headphones cut off and sound poured from the television. It seems like the repetition a minor episode. Although, it didn't cause the final dissolution in either case it was a notable recurrence in both life situations, particularly because of the timeframe in which it occurred during the slow break-up of both relationships.

The important point is that the same thing caused the headphones to stop working in both instances – Infinite Presence by using its messenger force Unified–consciousness. This secondary conscious force made direct contact with the conscious entities within the headphones which resulted in their malfunctioning. Although this does sound ludicrous, consider the possibility. One powerful consciousness communicating with a lower level consciousness to produce a desired, directed result because it was in accord with the ongoing process set in motion by the universe as a warning.

Note how in the first instance prescience must be considered because the loss of power in the headphones was to be duplicated by a future event of the same type. Is this also not a cause created by an effect – the earlier loss of power to the headphones resulting from the second episode, later loss of power to the headphones.

Even though the headphones episodes were in themselves of minor consequence they were an important part of the overall situation. In this case, it was important, as in all similar cases, because it gave a timeline upon which the recurring event was taking place. In the first loss of Ed's lover, the headphone event took place during the later stages of the dissolution. So, too, in the recurrence with Joan. This was not coincidence.

The next important repeat of an episode was the presentation of a finger ring as a gift to Ed's female companion by her new lover. This same act took place in both of my friend's lost loves. The rings themselves even looked alike! Each woman received the gift of the ring from her new lover and blatantly wore it in the presence of my friend. On the first occasion with Sue, my friend was shocked and was told it was only a token of friendship. On the second occasion, Joan told Ed she'd bought the ring for herself. By this time, my friend recognized the similarities of episodes, began to realize a deeper purpose was involved, but also instinctively knew it was too late to affect the outcome. Unfortunately, by this stage in both events of lost love nothing could be done by Ed to change the situation.

When the end drew even nearer, on both occasions both men whom Ed's lovers had left him for began to "stalk" both women, eavesdrop on their phone conversations, and would fly into abusive drunken rages which the women catered to by purchasing him even more alcohol when he ordered them out to buy it. Both women performed the same task to placate the same type of abusive alcoholic, during these separate, unrelated events!

In addition, both women ended up being deeply addicted to narcotics.

But there are even more recurrences to note. In each case where my friend's lover departed, the last name of the new person she chose for her new mate was the type of surname that is commonly used as a first name: hence, Mr. David and Mr. George. Also, in each event, the woman involved attempted to set up a situation whereby she could continue in a relationship with both my friend Ed and the other man at the same time with each of the men living within the same building! Can these extreme similarities also be but coincidences?

Yes, they could have been. Both women may have had identical personality traits. And their new lovers could also both had the same personality traits. For all four people! And how likely is it that the exact situations would be replayed in the exact manner episode by episode? Then the idea of coincidence becomes untenable. As noted, there were several individuals involved in these occurrences. Could it have been just coincidence that each played the same roles as would have been expected if pre–designed repetition were not involved?

And these were only a few examples among dozens more of the same type of episodes that took place during both dissolutions of Ed's relationships!

Each step by step episode that occurred during the two events were precautions sent to both Ed and his female companion at the time. This is a basic example of the re-enactment of a life event with similar instances taking place within the whole. While the nature of the events themselves are out of our control, we can choose to react to them differently. This is basically the extent to what might be called free will.

These life events are repeated as warnings so that a person can react differently. It is a way that Infinite Presence is giving people a second, third or multiple chances. This is but one positive aspect of Infinite Presence. It may also be a way to prepare us for or provide a proof of the existence of alternative lives in the multi–verse. However, these re-enactments could also be interpreted as a form of pre–destination or even fatalism if one chose to ignore the precautions and not change behavior.

During the dissolution of his second relationship– the one with Joan – my friend received even more direct warnings, warnings upon which he did not act. One warning came when Ed found a diary entry he had written to himself ten years prior, during the occurrence of his first love lost. He found this while "randomly" searching through some papers in a desk. The ten–year–old entry described how miserable he felt at the time of the dissolution of his first relationship. Instead of reading it and taking a lesson from it which might have taught him that the same event was happening again with his second lover he chose to ignore it.

Then Ed found another important paper from that same earlier period that he'd written which chronicled the tactics he would use to mentally sustain himself after the first relationship had totally ended. He did not accept this as a warning either during the dissolution of his second relationship, and his second relationship ended in the same way as the first.

Were these warnings purposefully sent to my friend? If so – how?By direction from Infinite Presence. Unified–consciousness was used to impress on my friend the idea that he should search through his desk where he found those important documents of his past experiences. It was then he had a choice to either ignore their content, which he did, or take a lesson from them and apply it, which he did not.

Again, a person could say that the second dissolution would have happened naturally without any conscious design directing it. To this I agree. However, it would not have happened in the same way. There were dozens and dozens of steps through which this dissolution proceeded and they were an exact repetition of the first which was not naturally caused and could not have been produced by sheer chance. Chance would have been even less likely on probability scales than as if God Almighty himself had been in control. The second dissolution happened the way it did to provide a possibility of change.

The outcome of these events, however, depends on the reactions of those to whom they are happening. If one believes that the repetition of a life event is a warning so that he can react differently to it then it must be viewed as a positive phenomenon. If he accepts them as simply omens of an unchangeable event, then they are negative and merely presage doom. But a doom that he could at least recognize earlier in the affair and extricate himself from sooner to avoid as much pain as possible.

Could positive actions have been taken by Ed to prevent the dissolutions of relationships? Yes, latent memory could have been initiated. These could be applied to the failing relationship in hopes of re-establishing better feelings. This is how. Latent memories from better times during an existing relationship could have been revived and re-enacted which to assist in the return to the former period of a closer emotional bond between the participants.

This could be interpreted as an act of grace from God. Or it could be a more chemically induced re–activation of past events which inspired the minds of the individuals involved. This could be produced either consciously or through a more spiritually based mechanism.

Thus, the meaning of the re-enactment of events as produced by Infinite Presence is determined by the individual to whom it is recurring and as such can be either beneficial or negative based on his reaction. Thereby, Infinite Presence is ultimately controlling all situations but allowing a participant to exercise limited free will.

REPLAY EFFECT

As in quantum physics, the perceived outcome is determined by the observer of the event. Yet, at the same time – as noted above – it allows a certain degree of latitude by allowing the parties involved to have input into the outcomes of some life situations.

Stated in another way, from James Bathurst: "Principally, or almost entirely, is it connected with events, or things, the recurrence of which is, regarding one's own personal knowledge, but singular or alone. Should it have occurred *twice* or oftener, a mental *circuit* is formed, and the force arising from the consciousness of one's impression is transferred to the *memory* of that preceding the present, thus completing the circuit between two points, or poles"

In addition, "Any mental state which creates attention or interest merely…instead of being discharged…is retained and thereby remains a force. As such it must have an effect. Mind being an electro–magnetic apparatus, ideas necessarily produce results."

Thus, according to this idea, the recurrence of events is a basic electro–magnetic event and an exchange of energy in a back and forth manner. The initial event is retained in the memory and can be re–played at a future date. It is up to the individual to recognize this and make the necessary changes.

There is another vital purpose for the recurring of events. In many cases, they are a response to a former activity the individual engaged in. If it had been a destructive, negative activity so too will be the recurrence factor in the same measure and time frame. In other words, the action he performed will be performed upon him with the same emotional results of either pain, misery, or happiness and joy depending on the initial activity. And it will last the same amount of relative time. Some people call this karma.

The time involved is relative because the suffering, or joy experienced, by any individual is relative. A week's worth of intense suffering at the loss of a lover in a relationship for one person might be equivalent to a year to another person. Only the soul can judge the depth of personal pain or joy.

A major factor in this concept is that Infinite Presence controls all matter and activity by acting directly on the conscious bodies that constitute all of reality; and that is everything that exists. It cannot be diverted from fulfilling a planned outcome. But it clearly can be petitioned for help or to have wishes granted. Some might call this praying. We can communicate with it directly through our own thoughts which are part of the same consciousness. (More on this later). Remember, we are obviously conscious, too. One of the primary purposes of Infinite Presence is to keep order throughout the entirety of all realities - by all means necessary.

MATHEMATICS OF RECURRENCE

The recurrence of events follows immutable scientific laws. One is that for every action there is an opposite and equal reaction. Also, in the electro-magnetic field of study, it is a general law, that all generated forces, tend to return from where they originated, directly or indirectly. So, that whatever occurs to produce a repetition, the person from whom it proceeds will also be affected by the return wave (?) (James Bathurst). Additionally, all physical events that take place follow a repeating pattern of steps which science can measure and which is the basis for experimental verification. And more simply, and as certainly, the biblical counterpart is: "As ye sow, so shall ye reap." It has never been stated any better.

In science and nature repetition is common. In fact, it is basic to apparent reality. Consider the recurrence of decimals, by which sequences of digits can infinitely repeat (Wille 2010). As the example provided by Wille: 1 divided by 81 will return the answer of 0.012345679012345679 and be repeated in infinite fashion.

Then there are fractals. When observing their structures, it is seen that the same geometrical pattern is repeated at ever smaller scales which always copy a primary form. Their nature of infinite repetition was demonstrated by the Mandelbrot Set discovered in 1975 by the mathematician Benoit Mandelbrot.

These forms of repetition are both scientific and spiritual. Many people recognize that concept. However, not all realize or accept the conscious thought that occurs within each particle that creates these patterns. And few consider how this exact method of repetition of existence applies to lives, and events, and human and animal behavior and is under the direction of, if not pure mathematical necessity, then certainly an Infinite Presence with a conscious purpose.

Whatever action you perform – be it negative or positive – it will rebound upon you in the same way, both in a scientific sense and a spiritual sense. The same way does not necessarily mean in only physically the same way, but emotionally, too. Both the physical and emotional are believed by most people to be the product of electro–chemical effects in the human body.

ETERNAL RECURRENCE

This is a belief that the universe and all that it contains recurs in the same form and exists in the same way repeatedly throughout eternity. We will be born again into our same selves, live the same lives and end the same way, only to go through the exact same life again. Reincarnation is not involved in this system; we return precisely as we were in the previous life.

This seems redundant and pointless and clearly does not allow for the possibility of change. However, it does allow for a person to lead a somewhat carefree and guilt free existence, knowing that all actions have already been performed and accounted for without reward or punishment, leaving him only to act them out again without any concerns for morality.

Eternal recurrence was a primary belief in the philosophy of Friedrich Nietzsche. He, however, seemed to have borrowed most of it from a man named Louis Auguste Blanqui, who himself may have borrowed slightly from the ancient Greek Stoics.

Blanqui was a then little–known, although brilliant, French socialist and activist of the late 19th century who greatly expanded on the concept of eternal recurrence, almost anticipating the Many Worlds theory of quantum physics. In 1871 while held in a prison in Brittany as a political enemy, he wrote the classic work *Eternity by the Stars* where he proposed his theory. According to Blanqui, every action of every person on earth is being duplicated on numerous, finite worlds throughout the universe, only to be repeated and again through eternity. Instead of being limited to one earth, this repetition of life is extended to other identical earth worlds in the universe. The people on each world are the same ones who are on earth but occupying a different planet and potentially at a different time. There isn't any relationship to reincarnation.

Rather than using mathematical formula to expound this theory, Blanqui used verbal logic. He believed that since time is infinite and it contains a finite number of potential occurrences, there cannot be any other interpretation than that everything that has ever happened will repeat itself an infinite number of times. Thus, Blanqui described eternal recurrence from where he was confined in prison. "What I write at this moment in a cell of the Fort du Tareau I have written and shall write throughout all eternity—at a table, with a pen, clothed as I am now, in circumstances like these."

Déjà vu? The concept of Eternal recurrence could certainly account for cases of Déjà vu, that feeling of having experienced a certain moment before or having been in a specific location even though seemingly having never been there previously.

He further states, "Until now, the past has, for us, meant barbarism, whereas the future has signified progress, science, happiness, illusion! This past, on all our counterpart worlds, has seen the most brilliant civilizations disappear without leaving a trace, and they will continue to disappear without leaving a trace. The future will witness yet again, on billions of worlds, the ignorance, folly, and cruelty of our bygone eras!

"At bottom, this eternity of the human being among the stars is a melancholy thing, and this sequestering of kindred worlds by the inexorable barrier of space is even more sad. So many identical populations pass away without suspecting one another's existence! But no—this has finally been discovered ,in the nineteenth century. Yet who is inclined to believe it?"

Two more quotes from Blanqui follow.

"The number of our doubles is infinite in space and time. In all honesty, one could not demand more." True doppelgangers in multiple forms?

"At the present moment, the entire life of our planet, from its birth to its death, unfolds, day by day, on myriads of twin–globes."

Not only does this describe eternal recurrence, but as already noted, it also recalls the Many Worlds theory of Everett. However, Everett's worlds were infinite and independent of one another, not mere repetitions of one earth in multiple locations.

APOCATASTASIS

In religious terminology, apocatastasis is normally described as the earth returning to the primordial state in which it existed at its beginning. This occurs after all "sinners" – depending on whose viewpoint is followed – have either been dispatched for punishment or have been converted into friends, including Satan. Armageddon will have occurred, Jesus will have reigned for 1000 years, and then earth will be returned to its original state and another, new Genesis will be instituted.

While this is a Christian belief that dates to the church fathers, it can also be found in many other systems in other forms, particularly that of the ancient Mayans who until recently had become quite famous for predicting the end of the world as we know it in 2012 to be followed by a new Age. Even though this did not occur, the scientific aspect of this belief is important to note. The prediction was based on a precisely calculated calendar, which, while perhaps not perfect, may simply be in error by only a short period. Yes, the world may still come to an end in this or the next generation based on the Mayan computations.

Here, of course, is seen a Mayan spiritual belief system that is based on mathematics. In this case, the mathematics appear to have failed to provide an accurate prediction of an event.

The general doctrine espoused by the common form of apocatastasis is that all civilizations and history come to an end and that the world is returned to its original condition from which everything begins anew. It is similar to eternal recurrence but only vaguely. In the apocatastasis model life is not simply repeated over again, it starts from an entirely new beginning. Those who had existed previously had either been transported to Heaven or sent to Hell, depending on which version one chooses to believe.

But there is another version of the concept of apocatastasis. Few people seem to be aware of it, but I find it particularly intriguing and eventually scientifically verifiable.

It concerns the universe as being finite with everything within it orbiting around a central point, irrespective as to whether the cosmos is expanding. Time itself, however is fixed in place, stationary, and across the entire universe. All elements in existence pass through this field of time and eventually re–cross the same location. At this point, the same environment that existed during the initial passage through this section of time is passed through again. This results in the same type of sociological, historical, and environmental events taking place as did during the initial and subsequent crossings.

According to the ancient author Julius Firmicus Maternus (mid fourth century A.D.), also identified as Julius Firmicus Maternus Siculus, stated that what is called the great apocatastasis occurs every 300,000 years at which point one orbit of contained space will have taken place. At this point, the earth would be reborn and re–enter its pristine state through which it will develop once more and follow the same type of history that it previously passed through. But not an identical history as propounded by eternal recurrence.

It will reflect the time–bound environment through which it proceeds with new actors and new events but all within the same theme of that which has gone before. A period of expansive learning in the past age of time will produce the same type of period. A period of war if passed through during this time–bound environment will produce a period of war. Great leaders will arise when the earth passes through the previous period in time when past great leaders had arisen.

Although the universe is of course much larger than the amount of space that could be covered in a mere 300,000 normal years, this formula is still correct. It applies to the orbital period of our specific galaxy around a central point of gravity in the universe.

In accord with the primary thesis of this work, Firmicus notes the ever-present consciousness inherent in the universe. "For the Divine Mind is diffused throughout the whole body of the universe as in a circle, now outside, now inside, and rules and orders all things." (Mathesis, iii., page 23)

At the time, Julius Firmicus Maternus (Siculus), was a famous Roman astrologer and lawyer already known for his book *Mathesis* written around 354 A.D. Here he further factored down the period of the great apocatastasis into segments of 15,000 years. He developed the dating system for the 300,000 and 15,000 year segments of the great and the intermediate apocatastasis from his extremely detailed and exhaustive astrological calculations, using Mesopotamian astronomical observations with Greek mathematics to reach his conclusions. While it may be difficult now to know exactly what the earth was like 300,000 years ago, it is possible to peer back 15,000 years and arrive at a comparison between then and now. Is there any earth period that we know of from history to compare to life 15,000 years ago?

About 15,000 years ago the earth was becoming a warmer and wetter place, entering the late glacial interstadial period. There are cooler, drier winters and hotter, drier summers. Population was increasing and better nutrition allowed for women to bear more and healthier babies. People have gradually departed from a hunter/gatherer life and are now living in villages where close communities have developed. (*After the Ice*, by Steven Mithen).

Unfortunately, this does not provide us with too much of a comparison. The differences between our worlds at these separate periods are far too extreme to juxtapose. But what about 300,000 years ago? What was the world like then? Can it possibly be like anything like the earth of today?

But, as expected, this is a much more difficult comparison to make. Neandertal (German spelling) Man was in the ascendant then and was still in a somewhat barbarous state. A cynic might claim that is an accurate comparison to society today. I will leave that to the reader to decide.

But for my purposes, there isn't enough data to compare the earth that existed 300,000 years ago to that of any recent history. Even environmentally the data is scanty. Does this mean Firmicus was wrong? No. Just that there isn't enough data available by which to make a reasonable judgment.

I must confess to the reader that on non–scientific grounds, and purely as a matter of belief, I accept the apocatastasis theory as true. Even a scientist can escape the bounds of measurements and mathematics when there is a deeper inspiration influencing him.

Apocatastasis can also be experienced at any moment of any day since our galaxy is continually passing through past areas of time. This is why I accept the idea of apocatastasis, having experienced moments of times long past while walking through the world of today. For a superb musical description of this phenomenon I suggest a song called *"Time Passages"* performed by Al Stewart.

I once again return to James Bathurst for a personalized account of how apocatastasis can also be experienced on a day–to–day level: in short, physically passing through other moments of time. "I have, when passing certain places, been affected with the same thoughts, and labored under the same sentiments, at just the same spot, as on former occasions, and when conscious, both of being there and of the mental affection; but of which consciousness and knowledge of locality I was, in the re-occurring instance, quite ignorant until subsequent arrival at a place well known. This shows that both time and place are necessary adjuncts, and without both, one is insufficient. And to be a successful cause for his repetition, the original impression remains, in mentally conceived shape and form, as potential."

Firmicus was a famous astrologer. His charts were extremely detailed and meticulously calculated. Is it possible that there may indeed be truth in the idea that the stars in their position can control aspects of life on earth? Wouldn't that seem to be a likely result if our planet and the rest of the galaxy passed through time-bound locations of space where instances of bygone periods are repeated? It is a concept to consider.

AKASHIC RECORDS/ETERNAL MEMORY

Akashic records, as most people are aware, consist of all the knowledge that has ever existed in the minds of all the people who have ever lived. But is that true in only this universe or the multi-verse? According to this book, there exists a distinct and separate form of Akashic record in all universes. But are the records the same for each universe? Again, per this word, the answer is no. Everyone who lives in each alternate reality – even though of the same indivisible spirit – experiences different histories once he has entered the parallel reality of existence. It is generally believed that Edgar Cayce drew most of his information from

Akashic records. But the records that he would draw upon in an alternate universe would not be the same ones in this universe. Thus, there must be an infinite number of Akashic records to account for all the parallel yet distinct realities.

Is there any scientific evidence that Akashic records exist in any universe? Unfortunately, it is yet impossible to contact alternate realities to investigate their Akashic records. Of course, the documentation of Edgar Cayce's remarkable career should suffice for our universe and proof of their existence in our reality should be sufficient to assume the same type of mechanics exists in the parallel universes.

But, rather than relying on the work of Edgar Cayce, I feel that offering proof based on evidence supplied by the scientific field itself would be more in order to present to skeptics. Especially since the evidence was provided by findings made in the discipline which is my specialty: archaeology and is impossible to dispute. It involves a well-documented case where the minds of numerous individuals from many different ages past were drawn upon to acquire information about the remains of an archaeological site that had never been recorded and whose existence had never even been anticipated.

The account of this event was chronicled in the book *The Gate of Remembrance* written by Frederick Bligh Bond in 1918. It is a relation of the discovery of a previously totally unknown addition to the Glastonbury Abbey which came to be known as the Edgar Chapel. At the time that excavations were proceeding at this historic site there was not any indication that this chapel ever existed until information concerning its location was received by automatic writing.

The two principals involved in this account were the chief archaeologist Frederick Bond and his associate John Alleyne. During numerous séance like sittings, spanning years, Mr. Bond requested information from the spirits while his associate Mr. Alleyne recorded by writing the telepathically transmitted answers given from the beyond.

At first, the information supplied was basic general knowledge concerning the history of the site that was under excavation. Then the type of information became much more specific and original as the spirits offered descriptions of portions of the site that had not been discovered and that no one even suspected existed. There was not any physical indication that there was anything beneath the ground to be found at the spots where the spirits indicated. No digging had even been planned for that location. That, of course, abruptly was to change.

According to one of the sittings, taken down through automatic writing, the spirit, unidentified at this point, was directly involved with construction of the hidden, unknown structure. "I made that building. All that I didde anywhere is fannes. And under them, three faire windows of foure lights with transoms and little castelwork on the ramps thereof. And if ye digge in the wall of the navis, there is much fell in. Serche the great pier of the nave opposite the cutte: yt is full…but they threw therein the fragments of my capella, a canopy at the west, and at the central ones – faire canopy work, and in the midst a little one for Our Ladye, sylver guilt and very faire."

What was being described in detail were the plan of a chapel whose existence had at that point been completely unknown.

But all communication was not about archaeology. Much of it concerned spirituality and what the world and life beyond this plane were like.

One of the primary "memory souls" that was contacted was that of a man who lived about the 15th century A.D. His name is Johannes of Glaston and he speaks in a language that was used in older England. The following extract from one of his communications – transcribed through automatic writing – well describes the method of contact throughout the sittings. It was taken from SITTING LI, January 26, 1912.

"I dydde it not, God wot, not I! Why cling I to that which is not? It is I, and it is not I, but part of me which dwelleth in the past and is bound to that whych my carnal soul loved and called 'home' these many years. Yet I, Johannes, amm of many partes, and ye better parte doeth other things – Laus, Laus Deo! – only that part which remembereth clingeth like memory to what it seeth yet."

Thusly, a portion of the memory of Johannes remains in the realm of the past as if it is still in existence there, but also in the future and other places as well. Is this the true eternal now! If the moment a person is living is the eternal now, what becomes of that eternal moment when the next instant of his life occurs?

And then there is the amazingly revealing extract from the September 15, 1912 sitting with Johannes. His words were once again transcribed by automatic writing.

"But one waiteth, even Johannes, whose body, scattered to the winds of Heaven, once lay in the cemetery of the monks, hard by the east side of the chapel of St. Michael in the midst of the graveyard. What matter? He lives yet in Universal Memory, and speaks and acts through every channel in which the Universal Life flows.

"Yet, when he is himself, he speaks well, as he was wont in the rude times that are as yesterday."

It seems as if Johannes is speaking of himself being separate entities in separate alternative worlds, doesn't it? And, he seems to also have contact with his other selves, through the various channels he speaks of, who also exist in these other worlds.

While the transcriber, Mr. Alleyne, may not be able to contact these other realities, it appears that once a soul passes from this realm there is not any restriction to communicating with other realities. Thus, it may indeed be possible that these other parallel worlds can have some form of communication amongst one another once a person has been translated from our realm and into another one.

A prominent philosophy of spirituality states that all around us is an illusion and is merely a projection of the dream of a godlike figure. We can not only transcend this through awareness, but also share in that godhood and can see through the divine unreality.

Even should this concept be true; it does not negate the possibility for a dream world state existing in ALL other realities as such. Thus, the idea of parallel worlds and the multi–verse is operational in all belief systems.

One potential link between this world and alternative worlds might occur during "normal" dreaming. It may be possible that during dreams consciousness can travel among various levels of existence. The dreams – or even nightmares – that a person experiences may in truth be real events taking place in another alternative realm. The dreaming person may naturally be attracted to this realm due to the simple fact of the relationship between the consciousness shared by both. This, however, is on the outer edge of conjecture and needs a great deal more investigation.

Continuing with the archaeological excavation at Glastonbury, it is to be noted that Johannes was only one of many spirit entities who were contacted. And during the various sittings, each spiritual entity would seek help and advice from other spiritual entities who had lived in often distantly removed times. Some lived in the fifth century, others in the twelfth, and so on. Time at this point seemed to offer no barrier whatever. Yet the spirits were all able to communicate freely.

At one point, an attempt is made by one of the contacting spirits to offer information about the world beyond. What follows is from the sitting of December 4, 1916.

"Cosmic facts are everywhere, but not easily attained…

"…by assembling yourselves together and obtaining the information ye seek consciously or unconsciously. The result obtained is the same, but the word endures….

"…The material world is the screen between – the complex fabric of the simple weaving. The essential facts are eternal which (? move) in a circle, and to them that know the circle, somewhat will pass into all times, only ye see but little at a time. The center is the point on which all revolves, and ye, revolving, are conscious of the influence, but cannot know the radius…."

Next is the final piece of a lengthy message that was given in the last sitting by an unknown being who spoke of Johannes and others involved in supplying the directions that led to the discovery of the until then unknown chapel.

"It was lovely, and he knew it, but when ye ask, 'What was it like unto?' he cannot tell you. It was heavenly – so was the sunset – and the shadows on the mere – but he could not paint these nor reproduce them for you.

"Those others, the great and simple, are passed and gone to another field, and they remember not save when the love of Johannes compels their mind to some memory before forgotten.

"Then through his soul do they dimly speak, and Johannes, who understands not, is the link that binds you to them.

"Learn and understand.

"WE WHO ARE THE WATCHERS.

"farewell."

APPOINTED TIME OF DEATH

Why do we die when we die? What determines that last moment when the final breath is taken and the soul or the spirit leaves the body? It's a question that seems to have no answer but which just about everyone who has ever lived wants to know.

An answer to this was given by James Bathurst who has been highlighted several times already. However, his reply is a cynical one and is based on a particularly grim view of the universe and the role of Nature in it. But, it is an answer which does not evade the question as many spiritual answers do and it does have merit, if viewed from an especially negative perspective. In any case, many people do see death as a negative occurrence and might find some comfort in Mr. Bathurst's observation.

To do Mr. Bathurst justice, his response to the question of "Why do we die when we die?" will be offered entirely in context. But first, his basic remark.

"When found incapable of suffering, he (any person) is no longer worthy of life; therefore having through experience arrived at an unfit state to exist, by becoming unimpressionable, Nature no longer has any necessity for such life, it being impossible to generate pain."

What this means is that a person dies when nature can no longer torment him with any more pain because he has become oblivious to it. Thus, as such, the person is discarded as of no more use.

In general:

"We perceive then that life is essentially for the endurance of pain. If Sardou had destroyed himself, no more grief would have been possible; and although in the latter part of his existence, probably such being favorable in contrast with the past, pleasure may surpass grief, yet the latter would be considerable, comparatively."

The natural law is that the greatest amount of suffering shall be generated from the greatest number.

Bathurst's complete response concerning the appointed time of death is:

"It is a principal or property of atomic–consciousness in the process of differentiation to generate from any sensitive organism the greatest amount of suffering possible. Human life is based on this law. It is a material instinct of all animate organisms to exert every means within its power to exert every means within its power to support, and to the very last extremity cling to life. Were it not for this innate power, the human race would become extinct, the difficulties and disappointments being unendurable. But, great as one's love of life, the burden of grief is so extreme, that many anticipate death by violent means. With advancing years men become dull and careless, the racking and acutely poignant grief causing comparatively slight effects. When arriving at old age, one is insensible to pain, and then dies. Such is necessary and expectant. When found incapable of suffering, he is no longer worthy of life; therefore having through experience arrived at an unfit state to exist, by becoming unimpressionable, Nature no longer has any necessity for such life, it being impossible to generate pain."

In common language again (not to belabor the point):

Q. WHAT DETERMINES THE MOMENT WHEN A PERSON DIES?
A. When that person has reached the point whereby no additional suffering can be forced upon him by Nature, he will die.

There is a great irony about James Bathurst. Although much of his philosophy appears to be very dark and depressing, if its message is inverted, or understood as having an opposite meaning, it is often rather positive and hopeful. For example, if "When found incapable of suffering, he (any person) is no longer worthy of life; therefore having through experience arrived at an unfit state to exist, by becoming unimpressionable, Nature no longer has any necessity for such life, it being impossible to generate pain" could also be understood to say, "Death occurs at the moment when a person can no longer bear his pain and he is released from his suffering."

PART II

MULTI–VERSE OF EXISTENCES

Nano science and angels

WEIGHING ANGELS

The ultimate question of this work factors down to the nano science, the sub–microscopic particles which do the bidding of nature or the Infinite Presence as representing God. It has been shown that particles can travel back in time and at speeds faster than light on the quantum level. The debate is whether these particles have the power of consciousness and, if so, is this consciousness under the intelligent direction of Infinite Presence or are following mathematical instruction as calculated without emotion by Nature. Of course, the question takes on another facet if it is determined that Nature has been created by the Infinite Presence and as such it must be under its direction and endowed with consciousness.

How many angels can dance on the head of a pin? (The actual original question concerned the point of a pin, not the head). You have most likely heard this question and may even have been greatly annoyed by it. Why would any angels be dancing on the head (or even the point) of any pin? There has been debate about who first asked this question. The most likely answer is William Sclater, an English divine who lived between 1575 and 1626. His original question was, whether angels: "did occupie a place; and so, whether many might be in one place at one time; and how many might sit on a Needles point; and six hundred such like needlesse points." This question obviously has been pared down quite a bit over the ages.

The problem had been posed over the years as an exercise in logic and physics. Could two physical objects occupy that same space at the same time? But is an angel a physical object? Then, could two spiritual forms occupy the same space at the same time? That's a good question, isn't it?

This question was confronted by the great Catholic philosopher Thomas Aquinas. According to him, it is not possible for two distinct causes (angels or etc.) to each be the immediate cause of one and the same thing. If you think that is difficult to understand, consider what the Pauli Exclusion Principle says on the same subject: two or more identical fermions (particles with half–integer spin) cannot occupy the same quantum state within a quantum system simultaneously.

Add to this, that by scientific measurement it can be assured that a great many more dancing angels can fit onto the point of a needle than stationary angels due to the angular momentum of the dancing angels and one will understand that this is a problem that has aroused a great deal of scientific discussion. Not only that, but the type of dance that the angels can perform is restricted by the laws of quantum physics in that since the angels must perform their dance at near the speed of light this means that only dances which do not require a great deal of precision will be suitable for the them to perform; such as the Twist from the 60's which required only bodily gyration. So, the angels that dance on the points of pins are probably doing the Twist originated by Chubby Checker rather than the Waltz made famous by Johann Strauss.

So how many angels can dance on the tip of a pin? Physicists take note. The number arrived at is $8.6766*10ex49$. However, there is a great difficulty here which very few people overlook. There may not exist that many angels! Particularly per Thomas Aquinas who simply stated, "It would seem that the angels are not in great numbers."

No matter how many angels may or may not exist, how can an angel be weighed? Does an angel weigh anything? If an angel is completely spiritual the answer must be no. If it is composed of atoms, the answer must be yes. Has an angel ever be weighed, assuming of course that angels do exist?

In the conventional sense, I have never heard of an angel being weighed. However, in the sense that an angel has some substance – which could be submitted to the weighing process – the answer then must be yes, if one accepts the evidence of witnesses.

The Bible has many examples of angels who display actual physicality which effects material objects. These specific angels are not simply ethereal spirits who take illusory form. One of the most compelling biblical examples is found in Isaiah, 6: 1–7. In this account from the Bible, angels have caused doors to move and have carried hot coals in their hands.

Isaiah 6: 6: *Then flew one of the seraphim to me, having in his hand a burning coal which he had taken with tongs from the altar. And he touched my mouth and said: "Behold, this has touched your lips; your guilt is taken away, and your sin is forgiven."*

The weight of an angel is computed to be one bit. One bit is the smallest piece of data, simply meaning that it is enough information to define an angel as being an angel. And it is based on this that the above weight for how many angels can dance on the point of a needle was determined. What does a bit of information weigh? It isn't really the bit that has the weight, but in this case, it is the information that is being described – the angel – that is the source of the weight. This would be more in the form of energy which can be weighed, similar to atomic weight. In other words – weighing the energy emitted by any given angel.

And for our purposes, the question of the existence and nature of angels has great relevance. The implication of the question is that angels are infinitesimally tiny, in fact, so tiny as to be sub–atomic in nature. As messengers of data, angels can take many forms, including states of consciousness under the direction of the Infinite Presence. Being conscious, they may even have been endowed with the ability to assume shapes so that if their messages are to be carried to people they can appear to them in a physical form. But it will be a matter of consciousness contacting consciousness during the transmission of data of one type or another.

There may exist different species of angels. Religions have already created a hierarchy of angelic beings. Could this hierarchy in fact be a special combination of conscious entities which are differentiated by atomic weight and number so that they can perform specific functions? Why could they not be in this form rather than purely spiritual beings?

After all, many angels do some "heavy lifting" for God (Infinite Presence).

ANGELS OF THE PERIODIC TABLE

Can angels truly exist as physical energy which transmits information directly to the mind and who can assume transitory shape as well as use telekinetic power to cause objects to move? Consider the comparisons below. It is not meant to be a scientific theorem or a religious tenet but basically a concept for thoughtful speculation.

According to Saint Gregory of the Roman Catholic faith, there are 9 choirs of angels. Oddly enough, there are also 9 Periods in the Periodic Table of Elements. A **period** is one of the horizontal rows in the periodic table, all of whose elements have the same number of electron shells.

However, it must be pointed out that the elements for period 8 in the Table have as yet (of this writing) been discovered, and that Period 9 is for the moment only theorized. Although the potential number is 9 and that agrees with the concept of this book that a potential must eventually be realized. So, let us assume that a 9th level will be found someday.

Basically, certain angels are designated with certain responsibilities and are given specific powers with which to carry them out. In the periodic table, certain elements have specialized nature – gaseous, metallic, etc. – and they are gathered into groups, like the choirs of angels, and that they have certain effects upon the environment and uses in the universe(s).

When angels have appeared, they are often in radiant form, sometimes nebulous, and often in human shape. These forms could correspond to the elements from which they are composed. For example, the more dazzling angels would consist primarily of the brilliant element phosphorous. The more nebulous or cloud–like of angels might be made primary of the element helium. Those which are more transparent than most might be largely made of calcium and silicon. And angels which could alter their appearance would likely exist in a volatile state as found in mercury.

Some angels are proposed to possess great physical powers and act as defenders of the faith and of certain individuals. These more powerful beings would most likely be formed from the highly radioactive elements such as technetium (Tc–91) which has an enormous half–life of 4.21×10^6 years. And the commonest of angels, the Guardian Angels, would be comprised of the most common, yet noble element such a hydrogen.

Since each of these elements possess a different vibratory frequency this would allow the angelic forms to communicate with other conscious beings through different modes of contact as well as appear – or fail to appear – in different forms.

Does this then mean that any particle of phosphorus is an angel? No. What it means is that certain angels of a certain species are created with an intense concentration of that element's properties and uses the vibratory frequencies inherent in these properties by which to communicate either telepathically or physically. And that the physical attributes assigned to these beings by witnesses are purely illusory in order to make the entity that is involved understandable to the observer by applying his own mental interpretation to what he is seeing.

Another important benefit to adopting these elemental states is that their condition depends on the environment in which they are placed. Liquids can become solids, gases can vaporize, and this type of flexibility allows an angel the ability to enter communion with many forms of existence in many forms of environment.

At this point, a person might object to the idea that after death he might continue into the afterlife and become a strange hybrid element from the periodic table – assuming he would become an angel. This is a misconception about angelic origin which so many millions of people seem to accept. In the religious sense, all the angels that are to be created have already been created. These are special beings of a unique nature whose specific purpose is to serve the needs of the Supreme Being. People never become angels, despite how strongly greeting card companies try to convince us they can.

Even in the religious sense angels can be viewed as forms of potent energy which can communicate in various ways with the outside world through their unique vibratory frequencies. The way human beings see them and experience them is a product of our own interpretation of a phenomenon that in the ordinary world does not make sense to us.

I will close this section with an interesting sighting made of a group of angels by three Russian cosmonauts. In 1985 while engaged in performing medical experiments aboard the Soyuz 7 space station, Russian cosmonauts Leonid Kizim, Vladimir Soloviv and Oleg Atkov were overwhelmed by a brilliant flash of orange radiance that came from outside of their spacecraft.

According to their report – which had been kept secret until one of them defected – they saw "seven giant figures in the form of humans, but with wings and mist–like haloes as in the classic depiction of angels.

They appeared to be hundreds of feet tall with a wingspan as great as a jetliner." At this point in time, the cosmonauts had been orbiting earth for 155 days and, since the Soviet Union was still intact, were most likely atheists.

Science of the Apocalypse

NEWTON CALCULATES THE END TIMES

As many people are aware, most prophecies concerning the end of the world are based upon mathematical calculations. This would place them in the category of scientific formulae. In the Book of Daniel, the timeframe of history is used as a measure by which to determine when the last days will be upon us. The faithful, self–taught evangelist William Miller predicted October 22, 1844 as being the date for the coming of Christ (after revising this from an earlier date). Even St. John in writing *Revelation* used numbers freely in his prophecies, assigning the numeral 666 to the Beast, which was given to him by an angel. Most prophecies concerning the end time rely on the Book of Daniel for the source material both numeric and non–numeric.

One of the greatest scientists of the modern era – if not of all time – Sir Isaac Newton also devised a date for the end of the world. He also relied heavily upon the dating system of events as compiled by Daniel. Even though Newton warned against "date–setters" for the end times, he calculated the year of the apocalypse to be 2060 A.D. This date can be located in two locations in the Yahuda manuscript in Jerusalem. The date was also found on an otherwise insignificant scrap of paper among Newton's possessions.

Newton believed in biblical prophecy in all its forms as being divinely inspired. One of his least well known works was *"Observations upon the Prophecies of Daniel, and the Apocalypse of St. John"* in which he wrote that his exposition upon Bible prophecy would not be understood "until the time of the end" and that the wicked would never comprehend it. He also did not predict an Armageddon type of destruction of the earth but that the last days would primarily involve the second Coming of Christ and the installation of a millennium of the Lord's perfect rule.

How did Newton arrive at 2060 as the date for the world's end? Below is his explanation.

So, then the time times & half a time are 42 months or 1260 days or three years & an half, recconing twelve months to a yeare & 30 days to a month as was done in the Calendar of the primitive year. And the days of short lived Beasts being put for the years of lived [sic] kingdoms, the period of 1260 days, if dated from the complete conquest of the three kings A.C. 800, will end A.C. 2060. It may end later, but I see no reason for its ending sooner.

Since 1260 days translates into 1260 years, Newton added 1260 years to 800 A.D. to arrive at 2060 A.D. as the year when Christ would come again in glory. The year 800 was prominent because it was when Charlemagne revived the Roman Empire which was then renamed the Holy Roman Empire.

PROPHETS OF THE BIBLE

The prophet Daniel was one of the greatest of those who could see into the future. St. John who authored the Book of Revelation is revered as a true visionary. Isaiah was renowned for his accurate predictions about the first coming of Jesus. From where did they and the other great seers get their knowledge of the future?

The most widely accepted non-spiritual view on this subject is that most prophets of any type are experiencing the aftermath of some form of epileptic seizure brought on either by illness or drugs. Unfortunately, many of the people in the medical field then simply discount the visions as delusions or products of temporary madness.

But in this regard, it should also be noted that the ancients and so-called "primitive" cultures often viewed epileptics and other people of an unusual nature as being "blessed" or gifted with special powers of second sight.

Experiments have been conducted on individuals while the subjectswere in the very midst of experiencing visions or epiphanies. EEG and other readings show that the brain is in a highly agitated, even electrified state at the time. These same studies also usually attribute the visions that follow as being the aftereffects from some form of epileptic episode.

However, these medical professionals seem to overlook the possibility that the subject's supercharged brain might be in reaction to the power of contact with Infinite Presence, some other spiritual force, or simply the Divine. If a direct conscious communication were being made directly with a force of that might and magnitude could not the brain be exposed to an electro-magnetic type of power which would naturally cause a great shock to it?

And some saints and prophets may in fact have been truly mad. But maybe this madness was the cause of direct contact with a power that was beyond their brains to withstand, giving them supernormal visionary powers.

Returning to St. John and his Book of Revelation, an important question to ask is how did he know what the effects of a nuclear war would be? He clearly described them. People melting where they stood and their eyeballs falling out of their heads. Shadows being their only remains (as in Hiroshima). And other effects of radiation poisoning. These could not have been caused by any naturally occurring catastrophe. They are the results of atomic warfare. Again, science and the spiritual combine to prophecy the future. Yes, an angel told John most of his prophecy but was the angel a conglomeration of specific, conscious atoms as previously mentioned? But, then again, aren't we all?

THREE DAYS' DARKNESS

One of the least known but one of the most important of spiritually based prophesies of the end of the world is one that is primarily adhered to by the Catholic Faith. It is a vision of a future event that has been proclaimed by three of Catholicism's most important "Divines" along with several lesser known religious seers.

It involves the world being overwhelmed by three days of totally unfathomable darkness whose cause is purely supernatural – originating from God. The purpose is to cleanse the world and bring about a new beginning. In this sense, it is very similar to the original theory of apocatastasis, which was mentioned earlier in this work. If you recall, one version of apocatastasis claims that the world will pass through a period of cleansing after all the Ages of earth have been completed with the result being the arrival of a New genesis.

The prophecy of three days' darkness, however, is not only a spiritual teaching, it is also an event that can be firmly founded in scientific reality. This was not known until relatively recently when the discovery of gamma ray bursts was made by the Hubble Telescope in deep space. Because the three days' darkness was not supposed to be caused by clouds, smoke from volcanic eruption, dust in the air from a possible asteroid strike or any other conventional manner, its mystery was purely of biblical proportions.

What follows are the verbatim descriptions of the three days of darkness as told by the three "Divines" of the church who have foreseen the event.

Blessed Anna Maria Taigi (1769-1837)

"God will send two punishments: one will be in the form of wars, revolutions and other evils; it shall originate on earth. The other will be sent from Heaven. There shall come over the whole earth an intense darkness lasting three days and three nights. Nothing can be seen, and the air will be laden with pestilence which will claim mainly, but not only, the enemies of religion. It will be impossible to use any man-made lighting during this darkness, except blessed wax candles. He, who out of curiosity, opens his window to look out, or leaves his home, will fall dead on the spot. During these three days, people should remain in their homes, pray the Rosary and beg God for mercy."

Marie-Julie Jahenny (delivered on June 15, 1882)

The days will be beginning to increase (days get longer on Dec 22). It will not be at the height of summer nor during the longer days of the year (summer time), but when the days are still short (winter time). It will not be at the end of the year, but during the first months (of the year) that I shall give My clear warnings. That day of darkness and lightning will be the first that I shall send to convert the impious, and to see if a great number will return to Me, *before* the Great Storm (Chastisement) which will closely follow. (The darkness and lightning of) that day will not cover all of France, but a part of Brittany will be tried by it. (However) on the side on which is found the land of the mother of My Immaculate Mother (the land of St. Anne) will not be covered by darkness to come, up to your place (home of Marie-Julie) . . . All the rest will be in the most terrible fright. From one night to the next — one complete day —, the thunder will not cease to rumble. The fire from the lightnings will do a lot of damage, even in the closed homes where someone will be living in sin. My children, that first day (of chastisement) will not take away anything from the three others (the 3-Day Chastisement) already pointed out and described.

Venerable Elizabeth Canori-Mora (d. 1825)

"...*Innumerable* legions of demons shall overrun the earth and shall execute the orders of Divine Justice... Nothing on the earth shall be spared. After this frightful punishment I saw the heavens opening, and St. Peter comingdown again upon earth; he was vested in his pontifical robes, and surrounded by a great number of angels, who were chanting hymns in his honor, and they proclaimed him as sovereign of the earth. I saw also St.Paul descending upon the earth. By God's command, he traversed the earth and <u>chained the demons</u>, whom he brought before St. Peter, who *commanded* them to return into hell, whence they had come.

These were the stories that nuns in black habits used to teach during Lenten season to terrify children in parochial schools. What made the event so frightening is that there wasn't any physical or atmospheric explanation for how such a phenomenon could occur. Now there is.

Our sun could be instantly extinguished if it were struck by a gamma ray burst. It would be as if a light switch had been snapped off and the sun was turned off. Thus, an event which scientifically at one time seemed difficult to place belief in has suddenly become very possible. Even potential.

Might this be another case where spiritual inspiration had tapped into a future scientific knowledge to prophecy an event that could happen in the future? What consciousness did the minds of these "Divines" contact to receive such special information? Or is their access to this knowledge just another so-called coincidence? Occurring to all of the "Divines" at different periods?

The three "Divines" just quoted lived prior to the 20th century and had no knowledge of gamma ray bursts yet they described the effects of such a phenomenon. Of course, they didn't call the event a "gamma ray burst" because neither they nor their hearers would have known at the time what that meant. But they did describe the effects of such an occurrence, just as St. John described the effects of nuclear war in the Book of Revelation.

FEBRARY 4, 1962 – PLANETARY ALIGNMENT

This was another prophesied apocalypse based on astronomical data. On the above date the world was predicted to end by the most illustrious seers and mystics in the world. Why? Because on this day all the planets would align on one side of the sun. The effects were supposed to be catastrophic, causing the destruction of the earth in the process. While this did not occur on our timeline in this reality, it may have taken place on one or more of the alternative earths

DOOMSDAY CLOCK

An interesting form of a strictly scientific method for predicting the future occurs in what is known as the Doomsday Clock. It first came into being in 1947 and is the creation of members of the *Bulletin of Atomic Scientists* science and security Board. There is an actual clock which hangs in their Chicago office which shows the nearness that the earth is to destruction due to thermonuclear war. It has been updated to also include a future catastrophe due to global warming as well.

Midnight represents the occurrence of global destruction, and the first setting on the clock in 1947 was set at 7 minutes before midnight. The closest to midnight – or total destruction – that the clock has ever been placed was in 1953 when it was set at two minutes to midnight due to the highly competitive testing of nuclear devices by the United States and the then Soviet Union in that year. The farthest from midnight it has even been was in 1991 when it was set at 17 minutes from devastation due to the signing of the Strategic arms Reduction Treaty (START 1) between the United States and the Soviet Union. An additional reason for the lessening of worldwide tensions was the dissolution of the Soviet Union during that same year.

MAYAN CALENDAR

A combination of scientific method and a religious belief system regarding future predictions can be seen in the ancient Mayan calendar. Immediately prior to 2012 it was of high interest in "New Age" circles and among the populous in general due to the calendar's prediction that the world would end in December of 2012. Whetherthe emotion that many people felt when this prophesy was not fulfilled was disappointment could be debated, it could not be denied that the world did not end at the predicted time. Or did it?

Some of the believers in the prediction argued that the world had indeed ended on a spiritual level and had passed into a new age, or a new era. What many people fail to understand is that the date on which the Mayan calendar began was randomly chosen. It coincided with the date of the original Mayan's creation which, as in the Christian calendar, is not a date that can with absolute certainty be verified. Thus, to prophesy that the end of the world would occur exactly on December 21, 2012 could not have been considered a reliable date. It wasn't any more reliable than the date of October 22, 1844 as calculated by William Miller. Although Mr. Miller's date had a far sounder factual basis as it was formulated on historical biblical events.

Ironically many of the followers of William Miller claim to also believe that the world had truly ended on the prescribed date. They believe that On October 22, 1844, Jesus had made a second appearance in this world and caused a psychic change to the planet and effected a spiritual renewal.

THE BIBLE CODE

One of the more scientific methods used for predicting the future was developed by a concept that was used in the book called the *"Bible Code"* which was written in 1994 by Michael Drosnin. He used a process called ELS or equidistant letter sequences and applied it to certain books of the Bible. Using a complex computer program, this process created long strings of letters starting with any letter value – say "A" – and then selecting every n^{th} letter to be part of this string. Dates of significance began to appear.And when, during further testing, it was found that important names appeared in proximity to the dates more notice was taken. For example, the name Yitzak Rabin was found. When it was realized that this name was located very near the date in the string on which Mr. Rabin was assassinated the significance of the Bible code theory became even more striking.

100

After a large series of tests were performed it was found that so many similar matches had been made between not only names but critical historical events that it appeared that the Bible had indeed been coded to reveal the future. But only if these "predictions" were read in hindsight it seemed. The future events could only be verified after the fact, which showed that these were revealed in the Bible but they were not of any practical value.

Another argument against the Bible code is that if the same test is conducted on various non–religious works the same types of results could be produced but with non–religious subjects and of less historical importance. Once this had become recognized, interest in the Bible code faded and has almost disappeared. Another argument made by some people is that the code that was placed in the Bible wasn't the work of God but that of aliens or some other higher intelligence. This does not seem likely.

There are two other views a person may take on this matter. One insight is that the conscious energy was transmitted by some means into the minds of those people in the current era who were reading these prophetic passages which allowed them to recognize them for what they were for the first time. These would constitute revelations during the so–called final days, or the nearing End Times.

Another viewpoint is that the Almighty placed these codes in the Bible to teach humanity a vital lesson. While humans could – in hindsight – see that the future had been predicted but, unless the prophecies were received by the method that God chose for revelation, they would be frustrated in their attempts to know the future. In other words – listen to the duly consecrated prophets!

Also, another message to humanity may be to pay attention to the words of Jesus when he said that only the Father knew the time of the end, not even the Son. So why does anyone else believe they can predict the day when the world will end? Maybe the Bible code was a practical joke played by God.

Finally, even the apostle Saint Peter had something to add on the matter of the end of the world. But, he too, may have been in a sense mirroring what might have been a practical joke played on humanity by God. Saint Peter also wrote a gospel. In it he wrote that the last days of the world would begin being numbered when the last soul had been born, since God had allotted only a certain amount of souls into human existence. The question then becomes: how will it be possible to know when this last soul has entered the world?

Quantum Imaging

VISIONS

I present this as a preliminary study in the possibility that naturally produced quantum imagine may be the cause of some visions that have been recorded taking place.

Visions can be described in many ways. They are often characterized as being hallucinatory – not necessarily illusory – and are generally sightings made of people, places or things which have a profound effect on the viewer(s) and can at times be impressed on the environment. The "visions" are usually visual but can also be audible or tactile but, unlike true illusions, produce effects that can in some manner be measured, either by light meters, electro-magnetic detectors, audio registers, or any number of instruments designed to record physical changes in the surroundings. As important as these measuring devices are the eyewitness accounts.

However, for our purposes here, visions are going to be studied in a totally unique way – by subjecting them to examination as possible effects like those produced by quantum imaging. quantum imaging is a recently studied phenomenon that is closely connected to the concept of particle entanglement. As you may recall, particle entanglement is a situation that exists in which a sub-atomic particle is divided and separated where each of the particles will both react to a stimulus even though only one of them had been subjected to it. Thus, if one half of the divided particle was struck by a laser and set into a spinning motion, the other half of the particle would have the same reaction even though it does not have any contact

103

with the stimulating laser. That is a simple explanation for particle entanglement.

Quantum imaging is a photographic or visual procedure which uses the particle entanglement process to produce images transferred from photon particles that had never come in view of the object of which the picture is being taken. In this way, pictures can be taken of objects from around corners and from other inaccessible places. As such, the camera is never focused upon the subject at all. And it isn't done with mirrors.

This process is performed by using particle entanglement, transferring visual data from one of the divided particles to another of the divided particles and then to the camera which processes the data. The particle that passes through the object to be photographed is recombined with its original half and then discarded after having transferred its image to the other particle. The newly combined particle then enters the camera where the image of the no longer existing first particle is captured for reproduction. A very fancy way to take a photograph!

With this method of taking photographs through quantum mechanics an intriguing potential explanation for the occurrence of certain types of visions is suggested. It may be possible that the scenes involved in personal visionary experiences are of a quantum level nature and that only a few physically distinct individuals can decipher these video and audio projections. This of course relates directly to the concept of "second sight" which people who are sensitive to experiencing visions are said to possess. Some of the visions occur only within the mind or "the mind's eye."

It also relates back to a belief James Bathurst professed, concerning the ability of individuals with exceptionally developed brain functions having the potential of contacting worlds beyond the "common" person's understanding.

Consider also the visions that have been experienced of the Virgin Mary. Could some of these have been produced by a form of natural quantum imaging? Could they have been created by an unlikely, but consciously produced, optical re-arrangement of sub-atomic particles so as to give rise to images that only certain people had the optical ability to see?

The figure of the "Virgin Mary" in Zeitoun, Egypt appeared over the roof of a Coptic Church and seemed as if made of brilliant neon light. The sightings began spontaneously, lasted for almost two years, then vanished as abruptly as they had begun. In this case, millions of people witnessed the images and the quantum effect – if so it was - must have been of a more common optical nature. Something like a rainbow in the sky. But being projected from an unknown source.

The apparition of the Virgin Mary at Fatima in 1917 will be used as a primary example of possible phenomena produced by quantum imaging without the aid on instrumentality because it was so well recorded and its visual effects could be applied to similar visionary events in general. But, in addition to the visual effects, also of extreme importance were the audible occurrences as heard by Lucia – the young girl who was at the center of the event – in which prophecies were offered and later fulfilled.

Only one person, Lucia, could see the Virgin Mary at the Fatima event. Was this due to some form of quantum imaging which only Lucia could see? And she alone could "hear" the prophecies given by the beautiful lady, which potentially could be described as harmonics of a conscious nature which could instill messages in the minds of listeners (in this case, one listener). The ultimate question is who was the creator of these quantum images and harmonic messages? And where was the creator's location.

At any rate, the Fatima event continued for several months in the year 1917 and was to reach a conclusion and a climax in October of that year. Lucia was promised that a great miracle was to take place on the 13th day of that month. This promise was made in July. Thus, since a miraculous type of occurrence took place on the 13th of October, it cannot be denied that some conscious power had the ability to, if not predict the future, cause amazing phenomena to arise.

On October 13th of that year a crowd of about 50,00 people – depending on unreliable estimates - marched through an unrelenting downpour to the site of where the miracle was to occur, known as Cova da Iria in Portugal. The press also arrived, complete with a corps of photographers.

The miracle began when Lucia reported seeing a sphere of light arise from the hands of the lady of the vision. Could this have been ball lightning? The conditions were perfectly suited for its production. A lengthy period of heavy rain had recently stopped and the air was humid and electrically charged – in more ways than one! Could not the power of the spectator's minds have contributed to the event like the sights created in the air over Bismarck's estates a few decades earlier where the images of war and battle appeared in the sky (which will be covered shortly)?

The crowd began to experience unusual phenomena. Some saw the sun change color, spin in a weird way, and lower from the sky. Others saw the entire landscape change color as would be seen on an LSD trip.

Children in particular noted showers of flower petals falling from mid air. And some people, even believers, saw nothing out of the ordinary.

The primary feature of this event was the astonishing color change that swept over everything in the area. Highly charged, iridescent light of a neon type quality is one of the major effects accompanying the creation of quantum images.

These types of colorful displays are common to Marian sightings. There is one other odd similarity among many of them; they were airborne. In Zeitoun the vision occurred exclusively above a Coptic Church; in Fatima on a hilltop; in Knock, Ireland above a Catholic church, and so on. Could this be attributed to a peculiar angular interplay among ionized proton particles which may be reflecting images of beings who exist in other places – like mirages – or even other realms?

Or are these conscious images that are always in existence around us but can only rarely be seen due to an unknown interface of realities, resulting in the creation of quantum like imaging? More testing needs to be done on this topic.

Then there is the matter of the prophecies made by the "being" who appeared to Lucia at Fatima. They came in three parts. The first was a description of Hell. The second declared that World War I would end soon and that another greater war would be waged sometime during the reign of Pope Pius XI, between 1922 and 1939. And the third secret described the fiery and bloody end of the world at some time in the future.

Is this a case of harmonics from the future taking the form of mental images in a person's mind? Many of the detailed descriptions of the future in the three "secrets" came true and were quite detailed.

Why did Fatima happen when it happened? How did it happen? Many people claim it was a religious miracle. Maybe it was in a non-religious way. Miracles come in many forms and from many sources. But even if it is a phenomenon of Nature, that does not disqualify it as being miraculous.

MAGICAL SIGHTS

Unexplainable phenomena of a demonic or evil nature caused by a type of quantum imaging event could also account for any verified cases of witchcraft activity, aside from the behavior induced by using drugs and other hallucinogenic potions.

Quantum imaging may also be able to explain some of the astounding feats performed by yogi, healers, and some acclaimed magicians and "gifted" people. This is not to imply fraud or deceit; it is to imply that these persons may truly have had access to special powers, powers whose real nature they may not even have been aware of. This also can help explain genuine astral travel and bi-location – being the result of manipulation of quantum fields by mental or spiritual powers not fully explainable. But quantum imaging seems to offer an intriguing opportunity into revealing the true nature of such "miracles."

Here I must invoke the name of the great Harry Houdini. This beguiling illusionist, inscrutable magician, undaunted escape artist, and master performer of astonishing feats of physical strength and muscular suppleness did not claim the use of any special powers. He could accomplish the seemingly impossible through intense and rigorous physical training and an extraordinary will power. But nothing truly magical. Nothing beyond human potential if applied to its limits.

Yet, there were still people who claimed that Houdini possessed special arcane knowledge that only true adepts of secret supernatural laws could wield. Even close associates of his argued that Houdini had the ability to bi-locate, moving from one location to another by dissolving his bodily elements and then reforming them. In a similar fashion, other people who knew him said that he freed himself from manacles and other devices by simply causing his body mass to shrink to such a size that the encumbrances slipped away, freeing him without additional effort.

But Houdini did not accept ownership of any of these so-called powers. He never claimed to have the use of any supernatural or magical powers and claimed that he was only a magician whose ability to perform seemingly impossible feats rested with his extremely strict training regimen.

It must be understood that no matter who the philosopher, teacher (guru), or prophet is who is propounding a belief system none of them is divine and all of them are human. Neither the ascended master nor the man on the papal throne possess any more first-hand experience with the afterlife or the other world than any cashier at any department store or restaurant chain. This includes Houdini.

All the highest philosophies and the most profound explorations into worlds beyond or the so-called world of illusion before us now are theories and not proven facts. The same holds true for the ideas proposed in this book.

While quantum physics can present scientific findings, they can only be considered as probabilities with a potential for error. So, too, must be the view taken of any spiritual systems or theological doctrines. The nearest we can approach the truth is found in the sacred scriptures of most of the major religions and even these are subject to human interpretation (misinterpretation).

Prophecies and premonitions

DAILY MIND MIRACLES

Prophecies are certainly a major feature of religious and spiritual belief. It may surprise some people to know that prophecies also have a major role in the field of scientific research and discovery. Even the great Sir Isaac Newton use his mathematical skills to predict the date for the world's end.

Prophecy and premonition have been the province of the biblical and supernatural for thousands of years. The source of inspiration was usually attributed to God or godlike beings who bestowed special information about the future upon the select few chosen ones. But there are other more rationalistic, scientific ways, in which this information could have been acquired. This is not to discount the supernatural source; in fact, the supernatural source – Infinite Presence – may be one and the same with the scientific source.

Infinite Presence can also produce premonitions in various ways, often by simply allowing thoughts originating in the future to inspire one's present mind. A close friend of mine recently passed away. I had not thought of him or contacted him for years due to different life situations.

For unknown reasons, I began to think about him. My thoughts became more and more directed upon him. No anniversaries or notices about him appeared anywhere. My mind simply began to dwell upon him. This began one week before he passed away.

Coincidence? Not likely. I had been given a warning of his impending transition by Unified–consciousness, but was unable to fully decipher the message. These messages occur all the time to everyone. Some say they come by way of angels. Some by mental telepathy. Some by spirit force. And some by Infinite Presence. No matter how the messages are sent, it is important not to ignore them.

HARMONICS FROM THE FUTURE

From *Atomic–Consciousness*: Thomas Brown, in *Evangelical History* states that a Christian woman, whom he knew, had two presentiments of her end. One occurred six months before her death, and consisted of exquisite music, such as no words can describe, which was heard at ten o'clock one morning, when sitting in her chamber.

Four months after, when in the same spot, she again heard the music, at which she dropped her work and remained motionless. Two months subsequently, she kissed her children, and informed them life would not be long, and shortly afterwards died.

Some people are particularly acute to music and the prophetic messages that can be heard in it. Part of this ability may be due to the sensitivity of that person's mind to the sub-atomic vibrations that may be delivered by the physical elements that compose the basis of musical sound.

Harmonic sounds can also consist of consciousness. Why should not sound itself possess a form of consciousness on a level that we may not even be able to comprehend? The blending of words – lyrics – with a melody can increase the power and communicative ability of music. Why are certain chants so awe–inspiring? The ancient Hindu chants in their various forms. The early Christian chants like those developed by Ambrose and Gregory. There exist both the wonder of spirituality and the symmetry of mathematics within them, creating a cosmic conscious contact that can instill peace and inspiration.

The process of communication by thought waves called radionics as considered earlier in this work also should be noted here again. This time as a way of transmitting ideas of events from the future to people existing in preceding time periods. Recall how radionics is a process which converts thought patterns into electronic harmonic waves. But instead of merely being transmitted as simple phrases or ideas to a human receiver the medium of transmission could include the harmonics of music, touching upon another level of comprehension.

The acoustic-musician Daniel Wilson has experimented with the process of turning thought waves into musical versions of the mind and has successfully produced audios of startling quality.

VOICES FROM SPIRITS

The harmonic resonance of spiritual voices and angelic voices, can also be the basis for premonitions, including the following rare occurrence from the American Civil War. Two missionaries visited the battlefield after one of the conflicts, to "administer spiritual wants." Finding one, among other men, apparently dying, they directed his thoughts toward Jesus Christ, but, while they were speaking, the man said, "Hush!" Imagining that he disliked to hear salvation mentioned, this only induced them to speak more earnestly on the subject, beseeching him to implore forgiveness. But the missionaries were silenced again by the dying warrior's order, "Hush, hush!" Enquiring the reason of this expression, he informed them that an angel was calling the roll, and silence was necessary to hear if his own name was on the list. The missioners were astonished, but listened. Presently, a reply came from the dying man, "Here!" It was an answer to the angelic call. He told the missionaries his name was on the roll, and he immediately expired.

From *Atomic-Consciousness*, Victorien Sardou, a French dramatist, was subject in his youth to great privation and misery. Frequently, he knew not where to obtain food. On one occasion, he went to the riverside to commit suicide, when a supernatural voice informed him that if he were to refrain from carrying out his purpose a great future was certain. He therefore retraced his steps, wrote some works which pleased the people, and by that amassed a great fortune.

MECHANICAL CONSCIOUSNESS

Another potential form of communication could be found in the mechanical, mathematically synchronized sound of machined devices. Why could not some people be supernaturally audibly affected by as simple a device as an old-fashioned wind-up alarm clock? A good example of this is provided in *Atomic–consciousness*. "A friend of mine, on one occasion, while his father laid ill in bed, observed the clock stop. It was a very unusual thing, and he went to set it going. In a few seconds, it stopped again. This was the case until he tried three times. After the third occasion, it kept going. Exactly twelve hours later the man (the father) died."

This type of story is quite common and verifiable. Along with the predictive power of the workings of the clock, the appearance of number three in these events should be noted. Often, a disastrous affair occurs after the third attempt. This seems clearly related to the mystical power of the number three in many religions, particularly Christianity.

Other premonitions can be visual, created by photon particles traveling at a slightly accelerated speed so that they become visible before the event occurs. Of course, this is another example of effect preceding the cause. This also recalls the process of quantum imaging. Why could it not also reveal images from the future as well as from other inaccessible regions?

The following is a personal recollection of James Bathurst. "One night, I and a school–fellow saw a light come from a house we well knew, and take a direction on a certain road. A few days after, a corpse was conveyed from the house, and carried on the same road." A premonition?

The next matter concerns a little-known fact from American politics. In 1841 John Tyler succeeded William Henry Harrison as chief executive of the country upon the death in office of the president. Few people are aware that the death of William Harrison had been prophesied to Tyler by a close friend, Littleton Waller Tazewell many months prior to the event. President Tyler gave his friend credit for helping him prepare in advance to assume the office of the presidency, setting the precedent as being the first vice–president to succeed to the presidency upon the premature departure of an elected president.

COMBINED IMAGERY

Particular sequences of numbers can even have a form of conscious ability to contact the mind directly. In the year 1880, a monk named Brogio, who had prophetic power of foretelling lucky numbers, was requested by two men to give the numbers which would win the forthcoming lottery. Because he refused, they beat him to death. When dying, he said, "thirteen" and "twenty–nine" would win. This statement proved correct.

Jedidah Buton, of Elmton in Derbyshire, was noted for his extraordinary mathematical powers. He was also enabled to foretell the day of his death, which turned out to be perfectly true.

The above examples were of a personal and private nature. Anyone can have a premonition since Infinite Presence is in contact with all things and all beings. An important factor involved is for the individual to be aware of the possibility and to try to understand and decipher any messages received be they through sound, sight or any other of the senses. And the messages need not be of a world changing nature; they can be of the most common form. It is a matter of being in tune with reality, the universe around one which is alive with constant activity and can be used for one's benefit.

Premonitions can also be formed by the concentrated force of many minds who were destined to suffer terrible catastrophes together. They could issue a plea for help or a warning from the future. The following example not only demonstrates this but it also reveals another case where effect preceded a cause. On this occasion, not only was it a reversal of normal causality it was an event which foretold of a coming disaster which unfortunately was not averted because the phenomenon was not understood when it was revealed. But if it had already happened, could it have been averted?

The disaster was a horrific explosion in the Morfa coal mine in Wales, UK., which caused the deaths of many men. But well prior to the disaster, the sights and sounds of the coming tragedy had been observed and heard by many witnesses. Unfortunately, no action was taken to prevent the premonition from becoming realized. And this seems very strange for a land such as Wales where supernatural occurrences are usually seriously regarded.

What follows is from the local newspaper report, *Christian Herald* – March 20, 1890 – of the supernatural manifestations that prophesied the coming disaster.

"…a few days previous to the explosion, noises were heard as if men were working. Doors were being opened and shut in that part of the mine where really no persons were laboring, and from which, therefore, no human agency could have emanated.

"...several days before the accident happened, miners working in various directions, most distinctly heard shouting, warning voices, and cries. The local press certified that on one occasion when the cage was ascending, the occupant felt by that **undefinable consciousness**, (author's highlight) which frequently enables one to become aware, in a darkened room or other invisible space, of the presence or proximity of a second or other person. The man entered the cage at the shaft bottom, alone, and at arrival at the pit mouth, no one else was with him; yet, most distinctly, while being drawn up the dark shaft, he became susceptible to another's company.

"But the banksman, saw step from the cage, a man whom he had previously never observed, and, instead of accompanying the other, proceed to a certain shed. The shed was the one afterwards used as a temporary mortuary for the recovered bodies.

"...the shoutings and slammings of doors in unused parts of the pit denoted the exclamations and calls, as well as probably, the rush from one compartment or division to another, of the men for safety, when becoming aware of the explosion. The presence of a second man in the ascending cage, typified raising the dead. It was also proved on a previous occasion, similar manifestations preceded an explosion, men giving evidence now, who formerly escaped."

Thus, not only was the disaster that was covered in the above news– paper account predicted by supernatural activity beforehand, but there was a previous accident that had also been forewarned of by the same type of phenomenon.

This was not an isolated incident. On occasion, individuals are endowed with what is called "second sight." They can receive information from the future by way of Unified–consciousness, (transmitted by quantum imaging?) which is another way that Infinite Presence attempts to help humanity. Unfortunately, a person or persons cannot be forced to heed the warnings given.

The following is also from a newspaper account. It also involves a mine disaster. This time it occurred in Canada at the Springhill Mine. Warning was given, an attempt was made to respond to this warning, but the result, as in the previously related mine disaster, was tragedy.

"A dispatch dated February 24th states that Mr. Swift, superintendent of the mine, reports that on February 14th great uneasiness prevailed among the men, from the fact that an old woman called 'Mother Coo,' the Picton Prophetess, had foretold three weeks previously that a disaster was to occur. So, to allay the excitement caused by this prophecy the management ordered a committee to descend and examine the working. This was done only a few hours before the explosion took place, and they reported that it was perfectly safe, no gas being anywhere present. Some of the men, however, averred that 'Mother Coo' never foretold falsely, and therefore refused to go down, consequently their lives were saved. Over one hundred were killed." *Pall Mall Gazette,* Feb. 24th, 1890.

At this point, a person might respond that the warnings and premonitions were pointless – except in the last case for those men who refused to go down the shaft –because the disasters occurred anyway with great loss of life. Some might even cry, "God's will!" But the fault is not with God, or Infinite Presence, or whatever supreme mind one wishes to blame because the warning had been duly given. The fault – if that word might be used – belongs to those who chose to ignore the warning and thereby were cause of their own misery.

Next is an account that demonstrates how mass individual consciousness can unite to form visions and other phenomena that pre–warn of a potential reality which did not come to fruition. But since it had the potential to become real, it most likely did so in an alternative ongoing reality in a parallel universe.

Psychologists might term this mass hysteria. However, that description only considers the symptom. The true cause is created by the thoughts of many people that can combine to create reality from mere consciousness. What follows seems strangely similar to the visions seen at Fatima.

"On Prince Bismarck's estates in Northern Prussia, at a place called Laurenberg, supernatural phenomena are nightly seen or heard. The spirits rap vigorously, visions are observed, and flaming swords appear in the air. For many months, the European people have believed a war to be imminent between France and Germany, and the Prussians believe these things to portend the coming struggle." *The Graphic,* February, 1887.

But this war was avoided on this level of existence. Nonetheless, the warning had been given.

Alternative Existences

POSSIBILITY FACTOR

Alternative realities are those existences which are created based upon a choice that a person made or did not make. If his choice had been different, a different reality for him and his environment would come into existence. This can take place on an almost infinite scale. I prefer to call it a scale of infinity minus one.

Such a concept has already been proved to be possible in the already mentioned double slit quantum physics experiment. Here it was demonstrated that photons passing through a screen with two slits cut into it when streaming through the screen multiply into all the potential realities they could assume and create an overlay of histories on the other side of the screen.

In cosmological terms, alternative realities are those parallel existences that occur simultaneously with one another but cannot ever be in contact with one another. In each existence, each potential choice that is made creates yet another alternative reality. And each alternative reality exists on its own non–inclusive timeline, which means it has its own past, present, and future and these timelines cannot cross or interface with other alternative realities. Their number is potentially without limit. However, for this work, they are all finally re–united when the ultimate observer decides to converge them into one essence, this observer being the Infinite Presence, or however one wishes to describe the Eternal.

As already noted a spirit can be both "dead"' and "alive" at the same time but not in the same segment of reality. An individual spirit can seem to be divisible but is not truly so because it can occupy any number of alternative realities simultaneously as its own observer, in its own timeline, existing in a world that had been created due to a decision that the individual had made to cause a branching of existences.

An example may help explain this. Suppose a man were to travel 500 years into the future in the universe in which he resided. He visits the grave of his wife who has been physically dead for centuries. Her body is only dust in the ground. And next to her grave is his own. This man is both dead and alive at the same time. By the same token, he could return in time, get his wife, take her into the future and show her both of their graves and both would be dead and alive at the same time. Even if not in an alternative reality their spirits can exist simultaneously in both states.

In the event just described the man and wife who were alive travelled to a different time in the specific segment of reality, or timeline, in which they existed. They no longer existed in the future, but their spirits could visit it as observers. In this reality they physically were dead. But their spirits continued in existence since energy can never be destroyed.

Once again we see a relationship between Infinite Presence and modern Judeo–Christian thought. Are these alternate realities, these infinite dimensions, the many mansions of the Father's to which Jesus often referred? Jesus could have described them as alternative dimensional realities but few if anyone would have understood him then. Even today.

Another example of religious comparison is the Tibetan Book of the dead. In this holy book, it is described how after death, during a stage of the Bardo, the newly released soul is tempted toward various realities by a variety of attractive colors. The wise soul would choose to follow the Clear Light to gain ultimate release. The other alternate realities that are anticipated in the Many Worlds Theory may be represented in the Tibetan Book of the Dead by the various paths of colorful light. These will provide the choices the soul of the newly deceased will make. The Clear Light of release, in the quantum theory, is that radiance projected by the Infinite Presence or the Eternal (God) when he, as the ultimate Observer, coalesces all existences into one.

The words being used in the Tibetan Book of the dead and the Many Worlds theory are obviously different but they describe the same concept of afterlife and the search for return to the Infinite Presence. Two examples were give here of commonly accepted religious counterparts but almost all belief systems share the same concepts.

A critical factor concerned with the movement of spirit among different alternative realities is the idea of individuality. Can the same person, with the same unique identity, exist in different alternate realities? Is not spirit indivisible and as such make this an impossible state of being?

While spirit may be indivisible, as already noted it can also be multi-locational; in other words, existing in more than one place but at different times and different dimensions. This is a condition which has already been proven to be possible by individuals who in this current reality have demonstrated the ability to bi-locate. Many of them were saints.

Others were mystics and yogis. But this applies only to the one realm of existence they inhabit. They could not pass among alternative realities.

The relocating after death of the spirit among different alternative lives in parallel worlds is not reincarnation. The same individual identity is maintained with each change of existence and lives in different time vibrations.

One of the prime concerns of Infinite Presence is to maintain the integrity of ALL levels of reality, meaning ALL alternative existences.

Why do alternative realities even exist? One reason they exist is because of their own existence. Double talk? Not so. Because one form of reality exists, an alternative form must also exist as a potential for the different options that might have been taken by occupants of the first noted reality.

An option is just as viable as any other feature of existence. If a person had made a different choice, there would have been a different outcome and thus a different reality will develop which produces the alternate existence. A potential must be given an opportunity to exist. And this also gives the individual spirit another chance to exist and continue living. This refers to the concept of eternity Eternal existence among many worlds.

On a spiritual level, the idea of potential realities as well as personal identities occupying them was the subject of discussion by the Eastern teacher Maharaj (even though these are false identities by his standards). "…just as…all colors are caused by the same light, so do many experiencers (*individual identities – author's notation*) appear in the undivided and indivisible awareness, each separate memory, identical in essence. This essence is the root, the foundation, the timeless and spaceless '**possibility**' (author's emphasis) of all experience."

Potential outcomes of various realities were considered by Niels Bohr. He once noted, "It is similar to the Copenhagen Interpretation in quantum theory. Reality only exists as a potentiality until it is observed to exist. Then it does. Prior to observation it can exist in all possible states at once until forced to choose a potential existence." God or the Infinite Presence, may be the one to force that choice ultimately.

DÉJÀ VU CLUE

But is there any proof that alternative realities exist? One might also ask: is there any proof heaven exists? If one had never visited London he might wonder if London really exists. A case for alternative realities has been made in the double slit quantum experiment where all possible existences were selected by the photons in the test. Proofs of heaven come mainly from religious sources, such as the promises of an afterlife made by Jesus. In both cases, however, the ultimate answer comes down to faith. And, as noted earlier, alternative realities are simply other definitions of heaven and eternity.

There are some examples that might point to the existence of alternative realities. Déjà vu maybe one of the best proofs. Doppelgangers – your personal double – and people you recognize but have never met before are two other lesser proofs. These represent potential crossovers from other realities. Although this would have to occur by some highly unusual interface under the control of a universal power.

Déjà vu however, is most significant as to possibly glimpsing into other realities. There have been many attempts to explain Déjà vu but it seems that the most likely explanation is that it is a moment from a parallel, or alternative reality, that has somehow become entangled with one's current reality. In this other reality, the action you have or will take next would have been different from the one you are taking or will take in your current reality. This act will cause a shift in realities.

The action of Déjà vu may be that actual junction point which marks the location where a choice made – or a choice not made – had caused the creation of an alternative reality. It is a focal point in eternity!

But how can an action from a different reality be experienced since we cannot have true contact with those realities while existing in our current one? Such contact can occur because for an instant the aspects of these two alternate realities exist in an interface created for that purpose by the Infinite Presence. There is nothing beyond the control of Infinite Presence.

Some form of contact among alternate realities seems to have been demonstrated in the case of the Glastonbury communications. However, the exchange of information among alternate realities appears in this case seems to be possible only to consciousness that has past from "life" to "death."

Another seeming proof of the existence of alternative realities is the doppelganger. However, this would only apply to a doppelganger which is truly yourself from a parallel world not simply a person within your own reality who looks exactly like you. A genuine doppelganger in this instance would have to originate in a parallel world.

And, as observed by Louis Blanqui, "The number of our doubles is infinite in space and time. In all honesty, one could not demand more." This quote also applies to what follows.

Less rare are the observations of people who one recognizes as having been seen before but who cannot be identified. These may be images from alternative realities who your other self knew or knows there and who for some reason momentarily crossed dimensional planes in an interface provided by the Infinite Presence. This may occur between people who share a deep love relationship since love can transcend all. We cannot contact these other realms – with perhaps a few of the already noted possible exceptions – or without divine assistance and even that is more than miraculous.

Yet, there must be some point of contact or the alternative world could not have had an origin. Again, the occurrence of Déjà vu may be the connection point – the zone of contact - between alternative realities.

Timeline of Eternity

MANY WORLDS OF HUGH EVERETT

As had been noted earlier, one of the primary premises of this work is founded upon the principle discovered in the double–slit quantum physics experiment. This experiment demonstrated that photons when ejected through a screen onto a background target reproduce in a infinity of alternative states. This postulate was made by the brilliant quantum physicist Hugh Everett and, as already stated, is known as the Many Worlds Theory.

In addition, when the photons were subject to individualized observation by a special camera they altered their movement to follow the paths that had been theoretically predicted instead of their normal infinite routes. This act demonstrated the ability of conscious thought. The atomic structures were aware they were being observed and because of this altered their paths to correspond with the expected.

Hugh Everett proposed the Many Worlds Theory which arose from this double–slit test in 1957 while still a graduate student. His faculty advisor was the renowned quantum physicist Bryce Dewitt. Doctor DeWitt immediately recognized the uniqueness of this doctoral dissertation and noted how it was the most powerful statement produced in the field of quantum physics in the last 40 years. Yet the theory languished. Only silence greeted it from others in the field. This may have been because Everett was only a graduate student at the time. It may also have been because it was too advanced for others in the field to fully comprehend.

After graduating, Dr. Everett was offered a job at the Pentagon which he accepted. Ironically, his work there actually advanced the cause of worldwide peace. He spent his time at the Pentagon developing thermonuclear war scenarios, much like in the movie *War Games* and is the person in real life on whom the character in the movie, Stephen Falken, is based. Everett described in detail the potential causes and the results of every type of nuclear war event that could be expected, certainly using his knowledge of multi–verses as developed in his astounding Many Worlds Theory.

Due to Dr. Everett's work at the Pentagon, the idea of Mutually Assured Destruction (MAD) was formulated and adopted as an official policy. In other words, this work was largely responsible for keeping the world from blowing itself up during the Cold War based on both the multitude of likely causes that could start a nuclear holocaust and the specific results for each case that would surely follow. As in the movie, the only way to win a nuclear war was not to wage one.

Hugh Everett died at the age of 51, still unheralded. As another irony, the man whose theory provides the basis for the spiritual essence of eternity for this book theoretically was an atheist. His last wishes were to have his cremated ashes thrown out with the garbage and this was a wish with which his wife complied.

Why would such a multi-verse system on a spiritual level even exist? Because Infinite Presence (God), the First Mover Principle, would make it possible for all potentialities that could take place do so through the mechanism of the multi-verse. Thus, an unlimited number of worlds exists in the mind of the Creator to fulfill all possibilities of existence. Some people might be distressed that the Many Worlds doctrine produces paradoxes. Are they concerned that Infinite Presence would not be able to control and/or overcome any form of paradox? That should not pose a concern since Infinite Presence (God) created these so-called paradoxes in the first place and as such would certainly have the capacity to co-exist with them or alter them in any possible manner.

TIMELINES OF LIVES

An almost infinite number of alternative realities spring from each major decision that is made in a person's life on each life plane. What determines which major decision will create a parallel world is currently unknown.

A life plane – or timeline - is that timeframe of a person's existence in the universe he inhabits. The concept of the timeline has been mentioned earlier but it is so vital to the premise of this work that a detailed description of it is necessary. Each individual identity experiences one lifetime on the plane in which he lives. After dying, he is removed from that life plane. Death itself probably begins another timeline for that individual. But an infinite number of alternative identical individuals (the same self in different realities) also exists in other timelines (other lifetimes) in an infinite number of parallel universes.

Each specific timeline in each universe is the entire history of the world in which the individual lives though he only has one lifetime in it. The world's history on that timeline starts at its primordial beginning and concludes with however that specific history concludes, be it natural disaster, human created Armageddon or by simply running out of days as normal. But the person who had entered this timeline by whatever choice it was that had catapulted him into it is bound to that timeline. For example, a person reading this book now is bound to the timeline in which we are all now existing. We share the same history and ongoing lives.

If a person in this current timeline makes a choice that leads to a different reality that difference will not occur in this timeline but in an alternative one. For example, if a scientist who was living in 1941 Germany on our timeline gave Adolf Hitler the secret to the atom bomb this would not change the history of OUR timeline. That is because OUR timeline does not includ Hitler possessing an atomic bomb.

However, if this scientist had given Adolf Hitler access to a nuclear weapon a new, entire timeline on an alternative, parallel world would be created. From that point on, in that timeline, history would follow a different course. Would Hitler have used the bomb? Would it have been stolen from him? Would he have been assassinated? An infinite number of results could have come from Hitler being given access to the nuclear bomb in 1941. But NOT on OUR timeline.

The same type of alternative reality could take place regarding the US presidential election of 2016. In another parallel world, Hillary Clinton won that election. But this would not be part of OUR timeline, but of another "spin–off" alternative world.

This is of critical importance. The option or choice that was taken creates an entirely new timeline in an entirely different alternative world, not the world in which the original choice was taken.

Consider the alternate world again in which Hitler was given access to a nuclear bomb. If that timeline only begins to exist from that precise moment – from the precise moment when the secrets to a nuclear device were given to Hitler – from where does the previous history of this newly created timeline come? Where is the past for this alternative reality?

This is explained by another paradox of quantum physics. The answer owes a great deal to the work performed by Professor Kater Murch at Washington University. In his experimentation with the effects had upon photons, or light particles, placed inside of a microwave box he discovered that when they interacted with the circuit he could observe their quantum states at different stages in their evolution. What this revealed was the astounding possibility that the future conditions of a particle can change the condition of a particle in its own past.

What this essentially means is that it is possible that present day activity can be influenced by actions that had been taken in the future! As such, a state of existence that had been created in an alternative world by an enactment of choice could produce an entire past or history based on actions taken in the future of the newly created alternative world.

Thus, in an alternative world where Adolf Hitler was given access to a nuclear bomb, the effects which this had on that parallel world's future history created that timeline's past. And this concept applies to all alternative worlds that come into existence. They created their past based on events that take place in their future.

This then presents the ultimate question on this matter: what is the indicator, or connection zone, that signifies that a choice – an option – had been taken which produces the new reality? Does some tremendous flash occur? Is there a thundering explosion? What happens to announce that a significant choice has been made on the timeline in which you are living which has caused to spring into existence an entirely new reality? What is that zone of connection between your reality and the newly created one? How can a person recognize it? Déjà vu had been earlier offered as a noticeable junction point. But could there be other indicators that point to a specific moment in time when an action had been taken which would create an alternative reality? This question will be studied further shortly.

CREATING AN ALTERNATIVE REALITY

The following event is real. It is being used as an example of how an alternative reality can be created not only for one self but also for others who participated in a life–changing moment of decision. The name of the main character has been changed.

Six–year–old Mindy stood before her drunken, enraged mother who was in such an intoxicated state that she pressed heavily down upon the child's shoulder from behind for support. They were standing in the driveway outside of their dilapidated home and facing an automobile a few feet away from them in which were seated Mindy's two slightly older sisters, and their father. The back door of the car was opened, beckoning Mindy to join her sisters and flee the horrible situation in which she found herself, at the mercy of a drunken, barely sane parent. Mindy's sisters were being rescued from this condition by a father who had recently acquired a great deal of money and was offering them a much better, happier life.

Mindy wasn't physically being prevented from joining her sisters; but she was being emotionally extorted by her mother. Using the words, "If you leave me I will die!" the mother kept her six–year–old paralyzed in place. Not much more guilt could be placed upon a child of that age.

Who could bear the burden of being responsible for her own mother's death?

The car began to inch away from Mindy and her mother. The car's back door was still open. Mindy could have raised her leg and taken that simple step forward. She didn't. The car started forward in earnest, the back door slammed closed of its own volition, and Mindy's life of abject misery – which did plague her from then onward–was sealed.

Not only that, but everyone else who was involved in the drama also had their lives changed.

In that one instant when Mindy could have stepped forward into the waiting car a decision was made which shaped many universes. In the current timeline when she did not choose to get into the car her fate of misery was determined as was the fate of many other people who were to associate with her from that point onward. However, on another timeline Mindy did step into the back seat of the car and brought into existence a totally different type of life, not just for her but also for the others who were part of the scene and a later part of Mindy's life as well. How do we know this? After all, Mindy did not get into the car, did she?

Maybe not physically. What about psychically? Spiritually? Had she accepted that option in her mind, in her consciousness, this would have created an alternative reality or, as some would say, an alternative illusion that would also be experienced by the Divine or by the Great Architect?

Upon being interviewed on this matter, Mindy gave an incisive description of that exact moment when she faced the decision of whether to get into the back seat of the awaiting car. According to her, "Time stopped," at that moment. There was a sudden, momentary pause to all that existed around her as if the world had become a still life painting. This same description has also been given by other people who faced a critical life changing option. Is this a connection point between alternative realities?

On a different level, the following examples might be more immediate to the reader. Consider your emotional attachments to any one of the following historical events and, as an often-asked question: where were you at the moment of their occurrence.

The crisis of 9/11. Tragic death of Princess Diana. When you heard of Michael Jackson's death. Queen Elizabeth II's Coronation. The assassination of President Kennedy. The first election of Barack Obama. These were major events which shocked the psyche. What did that MOMENT feel like? Did time seem to stop? Did the world around you become as a still life painting?

On a more personal level, consider the following events and ask yourself the same question. Recall the moment when: you met the person you love, asked him or her to marry you, the birth of a first child, being hired for that dream job, being fired from a dream job, death of a loved one. These are life changing events. What is the one factor they all have in common?

The impact that was made upon your mind! And that impact generated a conscious and a subconscious response. And among these responses was the creation of an alternative ideation, a hope for another path or a better path. This is what creates an alternative reality and a parallel world. Consciousness can and does create reality. But because to us it is intangible we cannot experience it or live it in THIS TIMELINE. An alternate timeline has been created however and it is populated by all the people you know and you yourself – as yourself in another reality – as well.

Some people make the mistake of thinking that their other alternative life would necessarily be better than the one they are currently living on. It isn't automatically true that the alternative option you would have taken would have produced a better, more positive result. In fact, just the opposite may have happened. If you had taken THAT alternative option a terrible disaster might have ensued and your life on that other timeline might have become one of misery and despair. You might have ended confined in prison, or possibly on death row someday.

And some people choose suicide as a way out of a miserable life or an unendurable time in their life. This may not be the best option either, possibly thrusting you into an even worse situation.

Thus, do not automatically regret the choices you have made on this timeline. They may have been the best you could have made.

REVERSE DÉJÀ VU

These questions are asked again. What happens to signify that a significant choice has been made on the timeline in which you are living which has caused an alternative world to spring into existence? What is that junction point between your reality and the newly created one? How can a person recognize it? Maybe reversing the effect of déjà vu is a clue.

The concept of déjà vu had been considered previously, including its potential source as being from an alternative reality. Maybe déjà vu is a moment from a parallel, or alternative reality, that has somehow overlapped one's current reality.

But consider déjà vu occurring in the reverse fashion. Suppose that in this current timeline you remember having performed an action or seen an action performed which the world around does not have a record of. You believe you only thought you performed it or only thought you had seen it. But somehow you intuitively or instinctively know that it occurred. This type of reverse déjà vu has happened to a great many people. Could this bewildering phenomenon be that point of connection between the act you had taken in this reality and the choice that you opted not to take? Did this omission of action create an alternative reality?

Alternative Afterlives

PRIME THEORY

This is the theory that the author is propounding. Like all of the other theories proposed by other people it is an idea based on various assumptions and is not a definitive fact. However, it is also this author's belief that we have been give the ability by the Creator – no matter how he or it is identified – to develop our own realities and in this way, choose or design our own afterlives with his ultimate approval. Thus, the alternative afterlives theory as proposed follows.

On every plane of existence – during any timeline – any significant life changing choice that is made by an individual causes an alternative reality to be created which sets into motion a chain of events which anticipates that new reality. The past of this new reality is formed from actions that are taken in this timeline's future which have been modified into this role of being in the past: thus, casualty is reversed so that the past has been developed from the future.

In these new alternative realms as well as on the original timeline, new worlds will also be created with the enactments of life changing options that were either taken or not taken. The results will transpire in this new alternative world timeline as in the original version; the past being created by the future and etcetera. Each alternative reality is separated from the other and never make contact – except on rare possibilities as already noted – after that unique singularity point where the option that was either taken or not taken created the reality of this alternative world in the first instance.

An infinite number of worlds can co-exist in the same space but occurring on different vibratory frequencies. There will always be enough "space." Each life lived on these will be unique. Yet, everyone will possess the same spiritual identity because they will all exist at the same eternal moment but in a different state of being. You will be the same self in all realities but will have different life experiences.

At a period chosen by God, Infinite Presence, the Eternal Mind or Aristotle's Prime Mover – which will be the Ultimate Observer – all the alternative worlds in all their timelines will be measured for validity against one another, perfected, and then merged into a single reality to result in a final afterlife of perfection. Even the so-called disinterested Prime Mover would act as the Ultimate Observer whose role is to measure, make perfect and coalesce the worlds into one Pure State of being.

This theory has similarities to other viewpoints on the afterlife. One of the most important is the concept of the Ultimate Observer, which can also be described as God, the Supreme Intelligence, Infinite Presence, First Mover principle. Etc. It is also a primary concept in quantum physics. It is the observation made by the prime observer – usually the scientist in control of the project under study – who confers reality upon an activity along with any type of measurement being taken, which can be as simple as a "mental" notation being made.

Ironically, being weighed and measured is a famous concept from the Bible as found in the Book of Daniel. It concerns the handwriting on the wall: Mene, Tekel, Parsin. Daniel translated this for King Belshazzar to mean: God has numbered the days of your reign and brought it to an end; You have been weighed on the scales and found wanting; Your kingdom is divided and given to the Medes and Persians." The comparisons to the alternative afterlives theory seem clear, if not exact.

Eternity is a major feature of the afterlives theory. It provides for leading virtually infinite numbers of lifetimes with new ones being constantly created. But these are not reincarnated lives as already noted. They are alternative lives assumed by the same spiritual identity to fulfill all possibilities of existence. The ultimate question is if or when the divine Observer will choose to combine all afterlives into one – if ever.

Returning to a previous matter, any personal belief system must be considered valid for that individual, be what it may. The purpose of this work has been to demonstrate how similar, and sometimes identical, belief systems are and especially how they are often verified factually by experiments made through quantum physics. Is this truly a dream world we live in? Probably, in one sense or another. Does it truly matter? Probably not if we are subject to the interpretations we ourselves make of the reality around us as we are taking part in the ongoing illusion. Is it our illusion or one belonging to Infinite presence, or God? Does it matter? Does it matter why the conscious photon particles decide to disperse into an infinite array when not under observation or create an expected double row pattern when the camera is recording them?

ANTHROPOLOGICAL VIEW

My background is in anthropology and as such I have studied a wide range of human behavior and how it has developed over the millennia. When studying any belief system, it is vital that the culture from which that system originated must be considered.

In many popular spiritual systems of an Eastern origin, individual human identity (not to be confused with EGO) is commonly seen as an impediment to advancement. It should be noted that in many of the cultures which hold this view individuality is not a highly prized trait. As such, to consider diminishing its nature and sublimating it to an overall Divine Intelligence in which all form of personality is submerged and thus lost is equated with advancement. This is a process that a person studying this type of belief should keep in mind and use as a comparison to other concepts.

In the Christian belief system of the West, the individual identity is generally more highly valued. As such, heaven or paradise or wherever the final destination is hoped for, is usually viewed as a place or location where the individual human spirit retains its own identity and finds eternal satisfaction basking in the glory of the Almighty. In a brilliant work titled, *"Our Self After Death,"* written by the Anglican Minister Reverend Arthur Chambers the argument is made for the extreme importance of maintaining personal identity after transitioning from the earthly plane because only by retaining the memories of actions and things in the former life which brought the person joy could he then experience the same type of joy or happiness as a reward in the world to come.

This seems contrary of course to the beliefs of many of the spiritual and new Age systems. Attachments to things and persons of this world is something to rid oneself of, according to their views. But at the same time, doing so makes these former attachments even more important in a sense of reality once they can be viewed through the eyes of enlightenment. It is a paradox which usually goes unexplained. And just as often, the teacher and those who are his students while insisting that all attachments with a novice's significant other, friends and others in his acquaintance be severed they at the same time argue that the attachments to the teacher and his devotees be strengthened. Thus, one set of attachments is being replaced by another. This "logic" still escapes me.

Some great teachers have attempted to align the belief system of both the East and the West. The most successful of these was Yogananda who was a person of the highest moral order and who certainly achieved the highest level of advancement, outside of Jesus, in this world. The depth and scope of his work is awe inspiring and can only be contemplated with immense admiration.

And, of course, there was the introduction of Transcendental Deep Meditation into the West by the Maharishi Mahesh Yogi. This allowed millions to use a powerful but easy to apply mode of meditation which vastly improves daily life and can lead a practitioner into the wondrous realm beyond thought.

For those of us who have been life-long anthropologists this discipline is very much like a religion. We have studied humankind as he has existed throughout all the ages. We have seen him in his lowest form, in his degradation and can aspire to see him rise to a special glory in the animal world. But we always place humankind among his fellow beings on earth.

Humans are earth dwellers and have always been earth dwellers, per the physical records that we have.

Spiritually we have a higher form of consciousness than other animals and it can be developed to even higher levels; but we are not gods. God is the only god, but we do share in his nature. But it is his nature in which we share, not one which we develop on our own as some people claim to be able do. As can be seen, this is a pure mix of science with faith. We are part of the animal kingdom who can be specially blessed by the grace of God to develop our divine nature – which is in all things not just us – but not by our own exertions but through the infinite mercy of the Supreme Being Himself. Yet, we are members of the Ape family whether we like it or not. There isn't anything that sets us apart as being any more special than the hummingbird. Could we begin to live the life a hummingbird if we were faced with that demand? Could a hummingbird become human tomorrow? Can a man or a woman become God by his own struggles? Let's not make ourselves the strange god who we worship and put that false image before the true God. He knows the difference!

APPENDIX 1

CURIOUS OBSERVATIONS

AN ATHEIST'S CONVERSATION WITH GOD AND HIS VIEW OF THE WORLD
(The following observations were made by James Bathurst).

I said to God, "Men talk of God. Would to heaven he would show it by depriving me of my wretched life. I would have taken it myself a dozen times, but shrink with horror from all pain."

He (God) replied, "Life will soon end. Let the glass (hour) run out naturally." To which my response was, "How men can love life is incomprehensible to those of thought. One thing alone can explain it – undeveloped mind.

"Whenever my fatal moment comes, God will have benefitted to an extraordinary degree from my overflowing cup of agony. I am a total wreck."

AND

Many persons who have an aversion to, or are incapable of penetrating the more uninviting aspects of life, anticipate the end of wars and social struggles, which would be replaced by tranquility and peace. Should, however, that desired consummation be reality, Atomic–consciousness would have been prescient of such, and in the dark unfathomable past, "before time was," pre–arranged such causes, as would in the fullness of time appear as opposing and counteractive forces.

Therefore, when the sword has been turned into the ploughshare, an enemy still more powerful will be at the door in the form of *cold*, and following its trail, *exhaustion of the soil.*

In more favored climes than our desolate northern homes, in sunnier regions, there dwells, or will visit, until disappearing before the above congealing foe, swarms of locusts which devour every green thing; or should such visits be occasionally impracticable, *drought* is substituted, whereby in its withering grasp, *hunger* and *thirst* prove fatal. Lest there be men yet remaining who far and away beyond the misty veil of the visible, *see* or *hope* of something better or brighter, we say, if *hope* makes glad the heart, hope on, frail brother, hope! But alas, in vain; thy expectation cometh not, and forms but an idle dream. For if one's imaginative conception be ever so distant, what remaineth? Strife and struggle for the last growing blade; strength and agility for hap–hazard drop. Are such times welcome? Not to the thoughtful. Rather clamor they for the speedy day when deathly knell of wailing and tears form the last rite over man's extinction.

AND

Repeated from the section: APPOINTED TIME OF DEATH

We perceive then that life is essentially for the endurance of pain. If Sardou had destroyed himself, no more grief would have been possible; and although in the latter part of his existence, probably such being favorable in contrast with the past, pleasure may surpass grief, yet the latter would be considerable, comparatively.

The natural law is that the greatest amount of suffering shall be generated from the greatest number.

It is a principal or property of atomic–consciousness in the process of differentiation to generate from any sensitive organism the greatest amount of suffering possible. Human life is based on this law. It is a material instinct of all animate organism to exert every means within its power to exert every means within its power to support, and to the very last extremity cling to life. Were it not for this innate power, the human race would become extinct, the difficulties and disappointments being unendurable. But, great as one's love of life, the burden of grief is so extreme, that many anticipate death by violent means. With advancing years men become dull and careless, the racking and acutely poignant grief causing comparatively slight effects. When arriving at old age, one is insensible to pain, and then dies. Such is necessary and expectant. When found incapable of suffering, he is no longer worthy of life; therefore having through experience arrived at a fit state to exist, by becoming unimpressionable, Nature no longer has any necessity for such life, it being impossible to generate pain.

Q. WHAT DETERMINES THE MOMENT WHEN A PERSON DIES?

A. When that person has reached the point whereby no additional suffering can be forced upon him by Nature, he will die.

WONDERINGS

James Bathurst: On one occasion, when about five years of age, and so ill that my parents were watching momentarily for my death, I saw, while lying in my cot, the clear blue sky, as observed in the darkness, and remarkably conspicuous with stars. Months afterwards, when comparatively restored, I, one night, stopping in a country road, amid some trees, to gaze on the grand stellar view, beheld exactly the same scene as so much impressed me when previously near death.

ODDITIES OF LIFE TAKEN FROM OTHER SOURCES

PURSUED BY LIGHTNING

Alfred Greenwood, on the 6th of July, 1845, was with some other men in a room gazing on the lightning. Greenwood, with an oath, said, "I wish it would come through and strike us all." Instantly, a flash came and struck the speaker dead, but none of the others were injured.

In more modern times, Walter Summerford was struck twice in his life by lightning and survived both events. After he died and was buried, lightning struck his tombstone.

Roy Sullivan, a U.S. forest ranger, was struck seven times by lightning and survived each strike. He is in the Guinness Book of world Records.

In 1945 a Japanese citizen named Yamaguchi survived both nuclear bomb attacks. He first survived the dropping of the atomic bomb on his home town of Hiroshima. On august 9th of that same year, he had to be in Nagasaki on business. He arrived there just in time to experience the dropping of the second atomic bomb on Japan. Mr. Yamaguchi also survived the bomb that was dropped on this city, and lived to a respectable old age.

CHOOSING ONE'S OWN COFFIN

(James Bathurst): Differentiation (Negation) is ever on the aggressive to thwart and disappoint. Happiness can be turned into misery by simply using words. If someone remarks to a person how well he looks, differentiation can take-over and cause the opposite to occur and sickness to take over. One day a man was digging a grave and told a passerby that "someone might be digging one for me next week." The thought was captured by differentiation and the thing spoken of lightly did take place.

Also, a carpenter was overheard to remark upon a batch of prime wood being cut up "That would be good for my coffin." It did become used as his coffin.

APPENDIX 2

EXCERPT FROM MICROMEGAS BY VOLTAIRE

*(**Visitors to earth from a planet orbiting the star Siriusand an inhabitant of Saturn questioned a group of sages about the soul and how they would describe it.**)*

Then Micromegas (from Jupiter) said (to the sages): "Since you know so well what is outside yourselves, doubtless you know still better what is within you. Tell me what is the nature of your soul, and how you form ideas."

The philosophers spoke all at once as before, but this time all their opinions differed. The oldest quoted Aristotle, another pronounced the name of Descartes, this spoke of Malebranche, that of Leibnitz, and another again of Locke. The old Peripatetic said loudly and confidently: "The soul is an actuality and a rationality, in virtue of which it has the power to be what it is; as Aristotle expressly declares on page 633 of the Louvre edition of his works"; and he quoted the passage.

"I don't understand Greek very well," said the giant.

"Neither do I," said the mite of a philosopher.

"Why, then," inquired the Sirian, "do you quote the man you call Aristotle in that language?"

"Because," replied the sage, "it is right and proper to quote what we do not comprehend in a language we least understand."

The Cartesian interposed and said: "The soul is pure spirit, which receives in its mother's womb all metaphysical ideas, and which, on issuing thence, is obliged to go to school as it were, and learn afresh all it knew so well, and will never know again."

"It was hardly worthwhile, then," answered the eight–leagued giant, "for your soul to have been so learned in your mother's womb, if you were to become so ignorant by the time you have a beard on your chin. But what do you mean by spirit?"

"Why do you ask?" said the philosopher; "I have no idea of its meaning, except that it is said to be independent of matter."

"You know, at least, what matter is, I presume?"

"Perfectly well," replied the man. "For instance, this stone is gray, is of such and such a form, has three dimensions, has weight and divisibility."

"Very well," said the Sirian, "Now tell me, please, what this thing actually is which appears to you to be divisible, heavy, and of a gray color. You observe certain qualities; but are you acquainted with the intrinsic nature of the thing itself?"

"No," said the other.

"Then you do not know what matter is."

Thereupon Mr. Micromegas, addressing his question to another sage, whom the Saturnian held on his thumb, asked him what the soul was, and what it did.

"Nothing at all," said the disciple of Malebranche; "it is God who does everything for me; I see and do everything through Him; He it is who does all without my interference."

"Then you might just as well not exist," replied the sage of Sirius.

"And you, my friend," he said to a follower of Leibnitz, who was there, "what is your soul?"

"It is," answered he, "a hand which points to the hour while my body chimes, or, if you like, it is the soul which chimes, while my body points to the hour; or to put it another way, my soul is the mirror of the universe, and my body is its frame: that is all clear enough."

A little student of Locke was standing near; and when his opinion was at last asked: "I know nothing," said he, "of how I think, but I know I have never thought except on the suggestion of my senses. That there are immaterial and intelligent substances is not what I doubt; but that it is impossible for God to communicate the faculty of thought to matter is what I doubt very strongly. I adore the eternal Power, nor is it my part to limit its exercise; I assert nothing, I content myself with believing that more is possible than people think."

The creature of Sirius smiled; he did not deem the last speaker the least sagacious of the company; and, were it possible, the dwarf of Saturn would have clasped Locke's disciple in his arms.

But unluckily a little animalcule was there in a square cap, who silenced all the other philosophical mites, saying that he knew the whole secret, that it was all to be found in the "Summa" of St. Thomas Aquinas; he scanned the pair of celestial visitors from top to toe, and maintained that they and all their kind, their suns and stars, were made solely for man's benefit.

At this speech our two travelers tumbled over each other, choking with that inextinguishable laughter which, according to Homer, is the special privilege of the gods; their shoulders shook, and their bodies heaved up and down, till in those merry convulsions, the ship the Saturnian held on his palm fell into his breeches pocket. These two good people, after a long search, recovered it at last, and duly set to rights all that had been displaced.

The Saturnian once more took up the little mites, and Micromegas addressed them again with great kindness, though he was a little disgusted in the bottom of his heart at seeing such infinitely insignificant atoms so puffed up with pride. He promised to give them a rare book of philosophy, written in minute characters, for their special use, telling all that can be known of the ultimate essence of things, and he actually gave them the volume ere his departure. It was carried to Paris and laid before the Academy of Sciences; but when the old secretary came to open it, the pages were blank.

BIBLIOGRAPHY

Bathurst, James. "Atomic-Consciousness." Harris & Haddon. Whimple, Exeter, UK. 1892.

Bond, Frederick Bligh "The Gate of Remembrance." E.P. Dutton & Company, New York, 1919.

Braden, Gregg. "The Divine Matrix." Hay House, Inc. Carlsbad, CA. 2007

Byrne, Peter. "The Many Worlds of Hugh Everett." "Scientific American Magazine." October, 2008.

Chambers, Arthur. "Our Self After Death." George W. Jacobs & Co. Philadelphia, PA. 1902.

Everett, Hugh III. "The Theory of Universal Wave Function." Dissertation. 1957.

Genovese, Marco. "Real applications of quantum imaging." Journal of Optics, 18, 2016.

Maharaj, Nisargadatta. (translator – Frydman, Maurice.) "I am That." Chetana, (P.) Ltd. Bombay, India, 2006

Marsh, Leonard. "The Apocatastasis; or Progress Backwards." Goodrich, Chauncey., Burlington, VT., 1854.

Maternus, Julius Firmicus. "Mathesis (iii)" Rome. ca. 354 A.D.

Mithen, Steven. "After the Ice Age. Harvard University Press, Cambridge, Massachusetts, 2003.

Sanberg, Anders. "Quantum Gravity Treatment of the Angel Density Problem." "Improbable Research." Royal Institute of Technology. Stockholm, Sweden. 2001.

Sugrue, Thomas. "The Story of Edgar Cayce." A.R.E. Press. Virginia Beach, VA, 1996.

T. Aquinas, *Summa Theologiae*, vol. 52, no. 3, 1266.

Tolle, Eckhart. "A New Earth: Awakening to your life's purpose." Plume Books, a division of Penguin. New York City, 2005

Walia, Arjun. "Physicists send particles of Light into the past, proving time travel is possible." "Collective-Evolution." December, 2015.

Whitla, William, Sir. "Sir Isaac Newton's Daniel and the Apocalypse with an introductory study of the nature and the cause of unbelief, of miracles and prophecy." Murray. London. 1922.

Wilson, Daniel. "Atomic-Consciousness." *ForteanTimes Magazine* (341) June 2016.

Yogananda, Paramahansa. "The Second Coming of Christ, Volumes I and II." Self-Realization Fellowship. Los Angeles, 2004.

www.ingramcontent.com/pod-product-compliance
Lightning Source LLC
Chambersburg PA
CBHW030444300426
44112CB00009B/1157